The Physiology of
TREMATODES

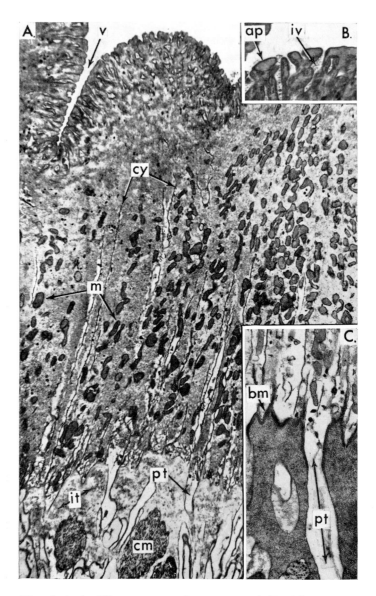

(Frontispiece)—Ultrastructure of tegument of *Fasciola hepatica.*
A×5500 External level of tegument (equals the ' cuticle ' of the
light microscope). B. Outer edge of A×14,000, showing invaginations. C×7500 Base of external level connecting with the cuticular
cells. ap, apical cuticular membrane; bm, basement membrane;
cm, circular muscle; m, mitochondria; it, interstitial material;
iv, invaginations of plasma membrane; pt, protoplasmic tube;
v, valley (courtesy Dr. L. T. Threadgold).

Trematodes

The Physiology of
TREMATODES

J. D. SMYTH

Professor of Zoology
Australian National University
Canberra

W. H. FREEMAN AND COMPANY
SAN FRANCISCO

Preface

The aim of this book is to give an account of the physiology of the Trematoda from the egg, through the developmental stages, to the adult worm. Although the Trematoda have been extensively studied, as far as the writer is aware, an account of the physiology of the group has not hitherto appeared in any language. This text is an attempt to produce a general picture of trematode physiology; it is not a detailed survey of all the physiological work which has been carried out on individual species.

Several difficulties arise in attempting to cover this field. First, most of the physiological studies have been carried out on species of trematodes of medical, veterinary or economic importance. Although this position is understandable, it does not necessarily follow that the species studied have been those most suitable for experimental work; indeed, this approach has undoubtedly resulted in the neglect of other trematode species which have special advantages for certain kinds of physiological investigations. The domination of parasitological research by pragmatic considerations reflects to some degree the historical development of the biological disciplines, at least until the last decade. Fortunately, the value of any parasitic organism as an experimental model *per se* is becoming increasingly appreciated and there is now a strong wave of interest in parasitism as a biological phenomenon, quite apart from any consideration of possible effects on the economy caused by the species of host or parasite.

A second difficulty is that many of the biochemical or biophysical techniques which have proved most valuable in studying physiological phenomena in free-living organisms, have been developed as precise tools only within recent years. Hence, so far as the parasitic platyhelminths are concerned, the data based on these methods are very limited. In some areas of the biology of the trematodes, then, much early work may have been carried out with methods or techniques for which more sophisticated techniques are now available. This is, of course, a common state

v

of affairs in scientific work and one which constantly confronts workers in every field.

It is now widely recognized that the advance of biological research has become so rapid and the amount of published work so vast that the ' half-life ' of any scientific text is inevitably shortening by the time it is published. This text is unlikely to be exceptional in this respect and advanced workers are therefore advised to supplement its contents by reference to recent published work, most of which can be found conveniently listed in Helminthological or Biological Abstracts.

J. D. SMYTH

Acknowledgements

I am grateful to a number of workers who have read, discussed and commented critically on various sections of this text: Dr C. Bryant, Dr J. A. Clegg, Dr T. Clubb, Dr D. Crampton, Dr K. Dixon, Mr B. Ewers, Dr D. W. Halton, Dr D. L. Lee, Dr J. Llewellyn, Dr J. C. Pearson, Dr S. R. Smithers, Dr L. T. Threadgold and Dr Winona B. Vernberg. Dr Dixon, Dr Threadgold and Dr Halton have generously allowed me to make extensive use of unpublished material.

Much of this book was prepared while I held an overseas Fellowship at Churchill College, Cambridge, and I am grateful to the Master and Governing Body for providing me with an opportunity to live and work under such pleasant conditions. My thanks are also due to Dr P. Tate of the Molteno Institute for providing me with laboratory facilities during this period and for much help given in innumerable ways.

The line illustrations are largely the work of Mrs A. Warrener who prepared them in Australia, and I am grateful to her for the meticulous care and patience with which she has carried out the task. My thanks are due also to Mrs E. Barta and to my wife, who each contributed several illustrations. A number of authors have also kindly contributed originals of illustrations or photographs from their papers. Most of the illustrations are redrawn from the original sources and acknowledgement is made to numerous authors, editors and publishers in this respect.

I am indebted to the following publishers or editors for permission to use material: Academic Press Inc., American Society of Parasitologists, Cambridge University Press, Springer-Verlag; Editors: Canadian Journal of Zoology, J. & A. Churchill Ltd., Compte Rendu des Séances de la Société de Biologie, The Company of Biologists, English Universities Press, Elsevier Publishing Co., Johns Hopkins Press, Liverpool University Press, Marine Biological Association, Masson et Cie., New York Academy of Science, Zoological Anzeiger.

Much of the typing of the MS. was done by Mrs E. Cowell (in Cambridge) and Mrs R. Rawlinson (in Australia) and I owe much to their skill and patience in setting out and preparing the MS. I am also indebted to Mrs M. Barton for considerable assistance with the bibliography.

Contents

SECTION I

PHYSIOLOGY OF THE DIGENEA

SECTION II

PHYSIOLOGY OF THE ASPIDOGASTREA AND MONOGENEA

To the Student

Scope of this Book

Traditionally, the Trematoda have been considered a class of the Phylum Platyhelminthes, the other classes being the Turbellaria and the Cestoda. The Trematoda were divided into three orders: the *Monogenea*—typically external parasites of cold-blooded vertebrates with direct life cycles; the *Aspidogastrea*—endoparasitic platyhelminths with the entire ventral surface forming an adhesive organ; and the *Digenea*—endoparasitic platyhelminths with simple organs of attachment and complex indirect life cycles. Studies carried out by Bychowsky[29] and Llewellyn,[195] based on a study of larval forms, have postulated that the Monogenea are more nearly related to the gyrocotylidean cestodes than to the trematodes. Thus, phylogenetically speaking, the Monogenea should be considered with the Cestoda rather than the Trematoda. Nevertheless, since they possess a well-developed alimentary canal and feed on solid or semi-solid host materials, their physiology is, on theoretical grounds, more likely to resemble that of digenetic trematodes than of cestodes. For this reason, a brief chapter on the Monogenea and the Aspidogastrea has been included, although surprisingly little appears to be known of the physiology of these groups. This volume is therefore mainly concerned with the physiology of the digenetic trematodes and throughout Section I, which forms the bulk of the text, the word ' trematode ', unless otherwise qualified, refers to a *digenetic* trematode.

Suggested Approach to Reading

The Trematoda are a remarkable group of organisms; study of them brings one in contact with the biology of a number of other groups, especially molluscs, arthropods and vertebrates which act as intermediate or definitive hosts. Furthermore, investigation of their physiology involves the use of biophysical, biochemical or immunological techniques as well as the routine cytochemical and histochemical techniques now widely used in biological laboratories. A proper study of even a single species can thus bring the student in contact with a wide range of scientific disciplines thereby providing an exciting introduction to fundamental biological research procedures.

Some knowledge of the basic morphology of a trematode is a necessary prerequisite to a study of its physiology, and morphological features of particular physiological interest are dealt with briefly in Chapters 1 and 2; and they are followed by a chapter dealing with feeding and digestion. These three should be read first. The metabolism of the adult is considered in Chapter 4, and this chapter, in which some knowledge of elementary biochemistry is assumed, could be left until the later chapters have been read. The latter deal with the biology of the egg and larval stages and, finally, development within the definitive host. The Monogenea are dealt with in Chapter 13, which should be read as a whole. The study of the relationship between a parasite and its host has developed into one of the most fascinating branches of parasitology, and the section on immunology (Chapters 10, 11 and 12) gives a brief introductory account of immunological principles before dealing with the specific problems of trematodes.

Finally, the chapter dealing with *in vitro* culture of trematodes, a technique which presents many challenging problems, can be read at any time, once a general understanding of the biology of a trematode has been obtained.

PHYSIOLOGY OF THE DIGENEA

1 : Digenetic Trematodes : General Considerations

General Account

Adult digenetic trematodes, with very rare exceptions, occur exclusively in vertebrate hosts where they are found typically in the major viscera such as the bile ducts, lungs and alimentary canal; one major group, the schistosomes, occurs exclusively in the blood system. Less favoured sites are the coelom, the urinogenital system, the swim bladder of fish, various sinuses or spaces and such aberrant sites as the eye.

A habitat which permits survival of an adult trematode must, in general, satisfy at least three criteria: (*a*) it must possess a connection with the outside world to enable the eggs to pass out of the body of the host (e.g. in faeces, sputum, blood or urine); (*b*) it must present a surface to which attachment by means of suckers is possible and on which feeding can take place; (*c*) it must possess an environment with a nutritional level sufficiently high to satisfy the enormous demands for energy and synthetic materials required for maturation and egg production.

With the exception of the alimentary canal, which is considered specifically (p. 22), it is not intended here to deal with the properties of the various habitats in which trematodes are encountered, but it is important that the properties of these should be borne in mind when the physiology of a particular species is being considered. As far as vertebrates are concerned, the nature of such habitats has been dealt with in some detail by several writers. [265, 290, 296, 344]

2

The digenetic trematodes differ from other groups of parasitic worms in that the first larval stages develop in intermediate hosts from the same Phylum, namely, the Mollusca. This extraordinary relationship between digenetic trematodes and molluscs is difficult to account for otherwise than by concluding that trematodes were originally parasites of molluscs and secondarily developed an association with vertebrate hosts. The parasitic phase may have been strictly confined to the larval stages of the ancestors, and the adult ancestor may have been free-living.[131]

This phylogenetic explanation receives some support from the fact that certain free-living flatworms, notably dalyellioid rhabdocoels, are commensal with molluscs and echinoderms and are clearly tending towards parasitism. The trematode redia also possesses certain rhabdocoel characteristics such as doliiform pharynx, sac-like intestine, paired protonephridia with separate pores and surface muscle sheaths. It can be easily visualized that when vertebrates evolved, the digenetic trematodes—perhaps in an encysted form—became ingested and adapted to vertebrate hosts (with the accompanying selective advantages of greater food resources and wider distribution) while retaining their connection with their original molluscan hosts in which they now undergo polyembryony. Although this hypothesis is the one generally put forward, the retention of the invertebrate phase in the trematode life cycle indicates that a molluscan intermediate host must have some selective advantage for the trematodes, an advantage probably related to the enormous reproductive capacity of this stage.

No clear ecological or physiological explanation can be put forward to account for the extremely complex larval stages formed in the Digenea. In addition to the molluscan intermediate host, a second intermediate host and, more rarely, a third intermediate host, as well as the definitive host, may be involved. This may result in a trematode being exposed to a range of habitats having widely different physico-chemical properties and exhibiting a range of nutritional levels. The genetical implications of this are considered later (p. 4). There is abundant evidence from many groups that parasites may undergo a diapause and may utilize the different properties of such environments to ' trigger off ' the next stage of development when a new habitat is reached. Thus, the egg of the blood fluke (*Schistosoma mansoni*) is inhibited from hatching within its human host by the high osmotic pressure of the body fluids, the high body temperature and the absence of light; when it is passed to the outside world it is stimulated to hatch rapidly by the lowering of osmotic pressure and temperature and the presence of light. Many similar examples can be quoted.

Basic Physiological Problems

Trematodes as experimental material

The trematodes as a whole provide superb material for many biological studies. A study of the detailed biology of a trematode involves the examination of at least three organisms—the parasite, its definitive host and its intermediate hosts. Quite apart from the anatomy of the hosts, which must be known in order to locate the parasite initially, it is also important to have data on host physiology and biochemistry since physiological or biochemical factors frequently play an important rôle in controlling trematode development or maturation.

Again, the actual structure of the boundaries of the habitat (such as the intestine) and the physico-chemical and nutritional characteristics of its environment frequently reflect the type of trematode parasites which could successfully adapt to these conditions; and to obtain information on this requires analysis by biochemical and cytochemical methods.

In analysing the host-parasite relationship, it is further necessary to have a knowledge of immunological reactions of the definitive hosts at a tissue and a humoral level, and investigation of these leads into the fields of immunology and serology (p. 184).

Genetics of the host-parasite relationship

The host-parasite relationship in trematodes is clearly a complex one involving (at least) three genetical systems; those of the parasite, the intermediate host and the definitive host. This may be schematically represented by fig. 1. The trematode must thus be suitably adapted over a wide spectrum of its characteristics—morphologically, biochemically, physiologically, immunologically and ecologically—in order to survive. Unfortunately, speciation in trematodes has been largely defined in morphological terms, but it is clear that all aspects of biology must be taken into account in order to obtain an integrated picture of its real phenotype.

Furthermore, trematodes have other characteristics (shared with some other invertebrates) which make it theoretically possible

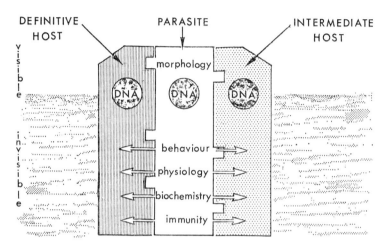

FIG. I. The host-parasite relationships of a trematode represented diagrammatically as an ' iceberg ' composed of three interacting genetical systems. Morphological adaptations are represented as being visible, whereas the—often complex—behavioural, physiological, biochemical and immunological adaptations between the parasite and its hosts, are not immediately discernible. The diagram could be applied to many other host-parasite systems. (original)

for them to produce 'strains' adapted to a new species or strain of host. These characteristics are: (a) they are hermaphrodites, and thus an unexpressed recessive mutant gene from one generation will appear in both male and female germ cells at the same time in the next generation; (b) self-fertilization may occur in some cases, or if cross-fertilization does occur, it is likely—on ecological and behavioural grounds—that mating individuals will have closely related (if not identical) genotypes. Thus, a *single* mutation could theoretically give rise to a homozygous double recessive and an ' instant ' mutant could appear. To these considerations may be added, (c) the fact that in the molluscan host trematodes multiply by polyembryony (i.e. multiplication of the original embryo), so that the mutant multiplies very rapidly.

These reproductive mechanisms could result in a large number of genetically identical individuals, i.e. a *clone* (in microbiological terms) developing from a single mutant. Thus a new ' strain ' of trematode, adapted to new ' strains ' of intermediate or definitive hosts and perhaps having distinct physiological characteristics, could arise. Chinese, Formosan and Japanese strains of *Schistosoma japonica*, for instance, are well known, although in this case, since the organisms are unisexual, the hermaphrodite factor outlined above does not apply. Although chromosome numbers are known for many species of trematodes, the physiological genetics of the group appear not to have been investigated. Nevertheless, as stressed above, the genetical mechanism for rapid adaptation to immunological, biochemical or physiological variations in the intermediate or definitive hosts clearly is present in trematodes and more extensive work may reveal the existence of more physiological strains than are at present reported. Apart from the complex patterns of host-parasite relationships, the adult parasite and its larval stages offer discrete, easily manageable material, which is particularly useful for studies in ultrastructure, cytochemistry, metabolism and biochemistry. Whole larval or adult trematodes, for example, offer exceptionally fine material for certain enzyme studies (plate I).

Problems of the life cycle

Some of the basic physiological problems which arise when the life cycle of trematodes is considered are shown in fig. 2. Data on many of these problems are meagre. The metabolism, for example, has been examined only in a limited number of species and yet a knowledge of it is fundamental to the whole of trematode biology, as well as being important in chemotherapeutic control of diseases of trematode origin. More, perhaps, is known of the chemistry of egg-shell formation and, since related structural proteins occur in many phyla, a comparative study of this process raises fundamental problems in biochemistry and biophysics.

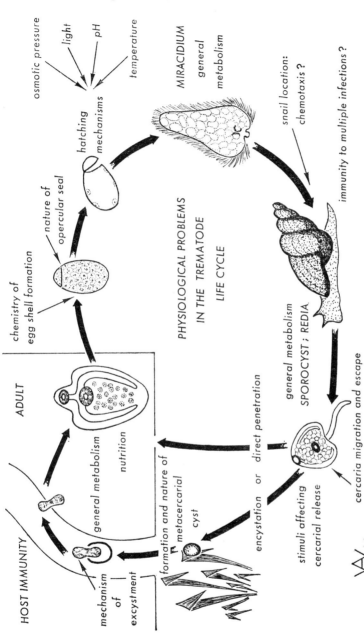

FIG. 2. A generalized diagram of a trematode life cycle based partly on *Fasciola* and *Schistosoma* to illustrate some of the physiological problems which arise. In many species a second, or more rarely, a third intermediate host is utilized. (original)

Again, the mechanisms of the hatching of the egg or meta-cercarial cyst provide elegant material for basic studies. In some species hatching of the egg appears to be induced by a light-stimulated enzyme (p. 69)—an unusual biophysical phenomenon whose fundamental basis has so far not been explored.

Passing to the larval stages, we find that little is accurately known of how miracidia find a snail host or what mechanisms govern the release of cercariae. Perhaps more is known of the cercaria than any other larval stage, a situation doubtless related to the ease with which specimens at this stage can be obtained; our knowledge of the behaviour of cercariae, however, is scanty. The metabolism of the larval stages, other than the cercaria, has been particularly neglected. Little too, is known of the nervous basis for trematode behaviour; and the discovery of neurosecretory cells in trematodes (p. 20) opens a particularly interesting field.

Finally, the whole question of host specificity and the tissue reactions of both intermediate and definitive hosts raise immunological problems of exceptional interest; many of these are closely interrelated with the general field of animal and human immunology.

This text discusses some of the problems outlined above, but it is clear that many aspects of trematode physiology remain unsolved, or even unexplored. Few invertebrate groups offer such a wide and interesting range of problems for further study.

2: The Adult Trematode: Some Special Organizational Features relevant to its Physiology

General Remarks

It is assumed that the reader is familiar with the general morphology, organization and life cycles of trematodes, a number of accounts of which are available;[61, 146, 290] these will not be dealt with in detail here. For reference, a generalized diagram illustrating trematode anatomy is shown in fig. 3.

It is clear, however, that certain features typical of the group are particularly relevant to a consideration of the physiology of the adult. Attention is especially drawn to the following characteristics:—

(*a*) With rare exceptions, all species possess at least one adhesive organ by the use of which they are brought into intimate contact with host tissues.

(*b*) Trematodes lack the tough, outer cuticle of nematodes; instead, the external covering, recently termed a 'tegument',[318] but often still referred to as 'cuticle', is homologous with that of cestodes and, like it, absorptive in nature (p. 13), at least in some species.

(*c*) Trematodes possess a well-developed alimentary canal, usually with a muscular pharynx. This type of system is especially suitable for ingesting semi-solid or viscous food such as intestinal contents, mucus, blood and bile.

(*d*) The reproductive system (with rare exceptions) is hermaphrodite and the eggs are generally protected by a tough, resistant
capsule.

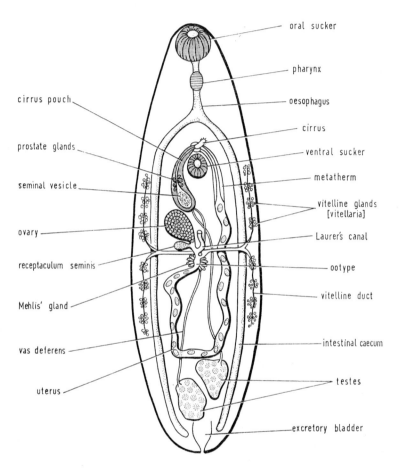

FIG. 3.   A generalized diagram of trematode anatomy.
(modified from Cable, 1949)

Some of these characteristics are discussed in further detail
below and attention is also drawn to other points of possible
physiological interest.

The External Covering: The 'Tegument'

In considering the relationship between a parasite and its host it is important to understand the nature of the external covering of the parasite, as this is the region which makes contact with the tissues of the host.

The trematodes have been generally described as being covered by a ' cuticle ' the homology and origin of which have long been a matter of dispute.[8] This external covering has been referred to by several other names, but the term ' tegument '[318] is used here.   Under light microscopy, the tegument appears as a continuous layer, about 15 μ thick with an affinity for basic dyes. The lower cortical layer appears strongly striated after certain stains and clearly consists of several layers, one of which is lipoprotein in nature (probably due to the basally situated mitochondria);   there is a well-developed basement membrane. Electron microscope studies[13, 28, 318] have shown that, as in cestodes, the tegument in trematodes is cytoplasmic in nature and is essentially a distal cytoplasmic extension of fusiform or flask-shaped cuticular cells which lie deep among (but differ from) the parenchymal cells.   The structure of the tegument, although essentially the same in all trematodes studied, varies from species to species and even in different parts of the body surface in the same species.   The basic ultrastructure of the tegument of *Fasciola* is shown diagrammatically in fig. 4.   The external or distal cytoplasmic region is more electron dense than the deeper or perinuclear protoplasmic region, and its surface is covered by a double membrane about 100 Å thick;   this membrane also covers the spines, which are thus intercellular organelles.   The surface configuration varies, being smooth in some areas and showing primary and secondary invaginations in others.   What appear to be microvilli within the secondary invaginations are probably just small evaginations of a very irregular surface.[319]   ' Pinocytotic vesicles ' have also been described at the surface membrane in *Haplometra* and *Haematoloechus*[28] but it is doubtful whether there are pinocytotic vesicles in *Fasciola*.   The lower part of the tegument is rich in mitochondria which measure only about 0·3–0·4 μ in width, and are considerably smaller than those found in the parenchymal cells.[13]   These appear in rows perpendicular to the surface

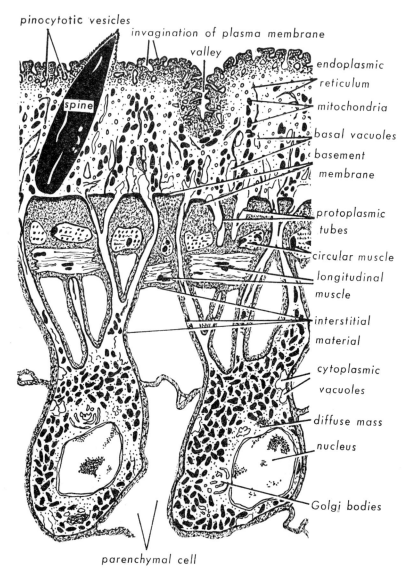

FIG. 4.   Ultrastructure of the tegument of *Fasciola hepatica*.
(after Threadgold, 1963)

and are chiefly responsible for the striated appearance of this part of the tegument by light microscopy. Rod-shaped bodies of high density measuring only 20–30 Å in section also occur between the rows of mitochondria; these were thought to be micromitochondria, but may be granules of synthetic materials (enzymes ?) passing up the protoplasmic tubes connecting the distal cytoplasmic region and the perinuclear cytoplasm;[28] this would seem to indicate a flow between the distal and perinuclear regions. These secretory products appear to synthesize in relationship with ribosomes of the endoplasmic reticulum; they may be collected and concentrated in scattered Golgi vesicles.[28] The mitochondria are somewhat atypical in that they contain few cristae;[13] their structure is more readily seen in the larger mitochondria found in the parenchyma (plate III, fig. C). This lack of cristae may be related to an anaerobic metabolism. The endoplasmic reticulum is tubular.

The ultrastructure of the tegument of *Fasciola* clearly suggests that, being a cytoplasmic structure, it should be capable of absorption, a view which experimental work has confirmed in this species, although much remains to be discovered regarding the mechanisms involved. Thus, if the mouth of a *Fasciola* is tied to prevent oral entry, glucose can still be absorbed,[211] a result which clearly points to active transport of glucose through the external covering. Again, specimens of *Fasciola* incubated in ferritin, absorb the iron particles at the tegument surface; these particles are then transported inwards, collect at the basement membrane and eventually reach the muscular coat.[13] This process has been called ' transmembranosis '. In *Fasciola* iron taken in with the blood can apparently be also excreted through the tegument (p. 24).

Sufficient studies have not been carried out on other species to indicate how far the structure of the tegument of *Fasciola* can be taken as general for trematodes. In the strigeid *Cyathocotyle bushiensis* the tegument covering the lateral and dorsal surface is thick, free from mitochondria and complex in structure and is quite different from that covering the holdfast organ. The latter is covered in a membrane elevated with numerous microvilli, which project outward towards the host surface. Associated with this microvillous surface are gland cells which show esterase and phosphatase activity (p. 15).[87, 88] Thus the tegument over the holdfast may function in the absorption of nutrients and the secretion of enzymes, whereas the tegument of the dorsal and lateral surface does not appear to be absorptive. It is clear then that the trematode tegument may vary in structure and function, depending on the habitat occupied in the host and the nutritional requirements of the parasite. In some species no absorption may take place through the tegument; in the strigeids absorption can occur through the holdfast organ and in *Fasciola*, and perhaps other species, absorption probably occurs over the whole tegument surface.

Organs of Attachment

General account

The intimacy with which a parasite can establish contact with the host tissue and the kind of surface to which it can become attached depend largely on the nature of the organs of attachment. These are generally well-developed in the trematodes and in some cases, as indicated above, can play a major rôle in digestion as well as attachment. Typically, a digenetic trematode possesses two suckers, an *oral* sucker surrounding the mouth and a *ventral* sucker, sometimes termed the *acetabulum*. Either or both suckers may be absent. The ventral sucker is generally the better developed. In the super-family Strigeoidea (the holostomes), an additional large adhesive organ, the *holdfast* is found behind the acetabulum, and as indicated later (p. 16) this organ is of special physiological significance. Some species of strigeids additionally possess a pair of *pseudosuckers*, one on either side of the oral sucker. Many species possess small unicellular glands which open anterior to the oral sucker. The function of these glands, which may occur in the metacercaria and the cercaria, is unknown. Most trematodes also possess small, unicellular, eosinophilic gland cells in the parenchyma which open ventrally in the forebody; their function is likewise unknown.

Suckers. These vary very little in structure. They are largely muscular in character with radial fibrils and equatorial and meridional fibrils near the rim, forming a sphincter. The oral sucker may bear spines and sensory papillae; being well supplied with tactile sensory endings (fig. 5). The latter may also occur on the ventral rim of the sucker. Gland cells in which esterases have been found[126] occur within the oral suckers of some species, e.g. *Haplometra*. This may indicate some degree of extracorporeal digestion.

The holdfast. The holdfast, sometimes known as the 'tribocytic organ' or simply as the 'adhesive organ' is confined to the super-family Strigeoidea. Its form is very variable but the commonest types are (*i*) the strigeid type, with two tongue-like

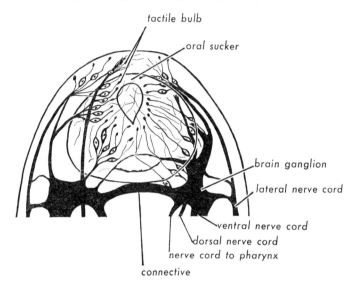

FIG. 5.   Innervation of oral sucker of trematode.   (after Bettendorf, 1897)

lobes and (*ii*) the diplostome type, with either a sucker-like organ or a single elongate protuberance.

The anatomy and cytochemistry of the holdfast have been examined in only a few species. [87, 89, 174] In *Cyathocotyle bushiensis*, a parasite of the duck caecum, the holdfast is acetabular and consists essentially of latero-ventral extensions of the body which fuse to form a cup-like cavity. There is no basement membrane, so that the ventral floor of the cup is, in fact, the ventral surface of the body.

The holdfast shows great physiological activity and is supplied with gland cells rich in RNA and enzymes (alkaline phosphatase, esterases and leucine aminopeptidase). These are secreted to the exterior and play an important rôle in digestion (p. 25). The presence of alkaline phosphatase in the walls of the microvilli of the holdfast of *C. bushiensis* suggests[88] that this organ, as postulated earlier, has a placental function and is capable of absorption as well as secretion.

Pseudosuckers. The strigeid pseudosuckers may act as quite powerful organs of attachment and in some species (e.g. *D. phoxini*) appear to play a major rôle in maintaining attachment to the gut wall. Like the holdfast, the pseudosuckers are richly supplied with gland cells which secrete esterases.[174]

Alimentary Canal

The alimentary canal is well developed, consisting essentially of an oesophagus, usually a powerful pharynx and a pair of intestinal *caeca* or *crura*; the pharynx may be rudimentary or absent.

It is generally held that the intestine ends blindly, but in fact a number of species are known to have anal pores.[225, 311] These can be grouped into four types: (*a*) caeca open into the excretory vesicle; (*b*) caeca joined to form a single canal opening ventrally; (*c*) caeca open to the exterior laterally, not reaching the end of the body; (*d*) caeca open at the posterior end of the body by two separate terminal anal pores.

The significance of anal pores is not clear. Possession of such structures presumably makes elimination of solid or semi-solid waste material easier and this may have a selective advantage in certain host locations. The histochemistry of the alimentary canal is discussed later (p. 26).

The detailed structure of the alimentary canal has been little studied except in *Fasciola*[69, 123, 301] and *Schistosoma*.[279] In *Fasciola*, the wall of the caeca consists of a single layer of epithelial cells which varies in height and possesses a basement membrane and a thin external layer of muscle fibres. These cells are presumably concerned with both secretion and absorption. Histological evidence suggests the existence of a secretory cycle, since the height of the epithelium appears to be related to the presence or absence of food material. When food is absent, or present in small quantities, the cells tend to be tall and columnar; when food is present the cells are short.   During the secretory cycle, the cells fill with secretion, collapse and liberate it into the gut lumen; the collapsed cells then regenerate.   Little appears to be known about the nature of this secretion, but probably it is chiefly enzymatic, since a number of enzymes have been detected in the intestinal caeca (p. 26).   This type of secretion recalls that found in active mammary gland cells, in which it has been termed *apocrine*.

The internal surface of the alimentary canal in both *Fasciola* and *Schistosoma* is greatly increased by internal protoplasmic processes which arise from the inner margins of the cells.[123] These processes vary in length, the longest being about the height of a cell.   The processes may be free or may form tubular loops; comparable loops have been found in sperm ducts and in the uterus.[327]   The protoplasmic loops may be essentially devices for bearing extended cell membranes, a mechanism which would permit greater molecular transference without reference to direction.

Excretory System

The trematode excretory system is a typical platyhelminth system with flame cells and collecting tubes.[290]   In general, it appears to contain no characteristic features distinguishing it from that of other platyhelminths.   Ultrastructure studies of this system have been made on the adult of *Schistosoma mansoni*,[279] the miracidium of *Fasciola hepatica*[170] and the cercaria of *Himasthla quissetensis*.[34]

A flame cell consists essentially of a main cytoplasmic portion with protoplasmic extensions at one end and a mass of cilia at the other.   On either side of the ciliary tuft the flame cell wall expands and forms a canal.   The wall in this region appears to be thinner than the remaining parts of the wall around the ciliary tuft and to be made up of alternating external and internal ribs.   The latter would provide a filtration mechanism resembling that of the glomerulus of vertebrates.[170]

The cilia are set in the flame cell cytoplasm in a repetitive hexagonal pattern similar to that seen in some hypotrichous ciliates, and the ultrastructure of an individual cilium resembles that described for other cilia.

A characteristic of the excretory tubules of trematodes is the

T.—B

presence of finger-like projections one to six micra in diameter.[327] These would greatly increase the internal surface of a tubule and their presence may be related to more efficient reabsorption.

The physiology of excretion in trematodes appears to have been almost entirely neglected. The system undoubtedly plays a part in osmotic pressure regulation, as it does in other helminths. In addition to the protonephridial system, a curious *paranephridial plexus* or '*reserve bladder*' system, is present in a number of families; it is well seen in a strigeid metacercaria. It consists essentially of a plexus of vessels connected to the protonephridial system which sends out ramifications to all organs. The paranephridial system contains calcareous corpuscles and small fat globules. When metacercariae of *Codonocephalus* are incubated in balanced saline at 37° C., the contents of the paranephridial plexus disappear within a few hours.

This has led to the suggestion[258] that in the metacercaria the paranephridial system serves as a store for higher fatty acids which result from the anaerobic breakdown of glycogen (p. 39). When the body temperature is raised to 37° C. these higher fatty acids are believed to be metabolized to lower fatty acids, which can be readily secreted, and so further energy becomes available.

'Lymphatic' System

In a number of families of the Digenea, e.g. the Paramphistomidae, there occurs a system of mesenchymal vessels which gives off branches to the viscera.[316] These 'lymph' channels are said to contain free cells not unlike primitive vertebrate blood cells (haemocytoblasts), and the whole may represent a primitive circulatory system. In some cases, this so-called 'lymphatic system' is derived from the primary excretory tubules. Nothing appears to be known concerning this system, which may be important to the physiology of a trematode.

Reproductive System

The male system follows the typical platyhelminth model and presents no unusual features. The female system is of especial interest, however, as the type of egg may be related to the metabolism and/or nutrition of the worm as well as to the environment within the host. The biochemical, histochemical and morphological problems of egg formation are closely interrelated, and for this reason the morphology of the female reproductive system is dealt with when these questions are discussed later (Chapter 5).

Muscular System

The physiology of trematode muscle does not appear to have been examined by the classical methods used for vertebrate muscle. The structure of the muscles of *Codonocephalus* has been examined by means of repeated injection of methylene blue.[258] The muscle consists of elements belonging to the usual type of platyhelminth muscle cell, having myoblasts attached to the edges of the fibres. These fibres are straight while at rest, but show a remarkable spiralization on contraction; this spiralization disappears completely when the fibres return to a state of rest. The observed effect is brought about by the squeezing together of a non-contractile sheathing membrane which may correspond to the sarcolemma of muscle cells in higher animals.

Nervous System

General account

The general morphology of the trematode nervous system—which basically resembles that of Turbellaria—is well known and has been described in a number of texts.[7, 66, 146] The nervous system in *Codonocephalus* and *Opisthorchis* has recently been described in detail.[165, 258] Essentially, it consists of a pair of cerebral ganglia and a sub-muscular plexus concentrated into longitudinal cords and transverse connectives. Typically, there are three pairs of longitudinal cords—dorsal, lateral and ventral; the transverse connectives may be numerous, 40–70 in *Codonocephalus*.[258] In some species (e.g. *Codonocephalus*), related to the general morphology, the nervous system is well developed in the region of the forebody but the cerebral ganglia are reduced. In this respect, this species resembles the land planarians. A clear nerve ring around the pharynx is considered to represent an autonomic nervous system, and fine nerves to the intestine and excretory system represent sympathetic nerves.[258] Motor and sensory nerve cells in trematodes have been described; the latter abound, especially in the suckers.

Neurosecretory cells

Neurosecretory cells are well known in the Invertebrata, especially in the Arthropoda, Mollusca, Turbellaria and Annelida. It is not surprising, therefore, to find that they have been reported from the Trematoda.[328]   In *Dicrocoelium lanceatum*, a pair of neurosecretory cells are present in the cerebral ganglion (fig. 6). The evidence that they are secretory is based on their staining

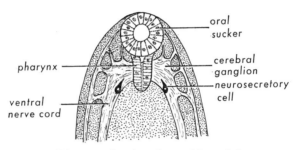

FIG. 6.   Diagram showing the position of the neuro-secretory cells in *Dicrocoelium lanceatum*. (after Ude, 1962)

reactions with paraldehyde-fuchsin and chromalum-haematoxylin-phloxin, the same reactions being observed also in other invertebrates. These cells contain secretion in the form of fine granules which can also be observed in other parts of the nervous system, especially the ventral nerve cord.

Cholinesterases in Trematoda

Cholinesterase activity has been detected both chemically and histochemically in the adult and miracidium of *S. mansoni*,[24, 246] in adult *F. hepatica*[344] and in the metacercaria of *Diplostomum phoxini*.[174]   *S. mansoni* possesses an acetylcholinesterase which shows substrate specificity and behaves, so far as substrate concentration is concerned, like the enzyme which plays an important rôle in neuromuscular transmission in vertebrates. Although little work had been done on the neuromuscular physiology of trematodes, recent histochemical methods for cholinesterase, using selective inhibitors, may lead to more rapid advances being made in this somewhat neglected field.

3 : The Adult—Feeding and Digestion

General Considerations

Investigation of trematode feeding mechanisms presents a particularly difficult problem, because of the location of the parasite within the alimentary canal or viscera of the host. This means that feeding *in situ* cannot be observed *directly*, so that information must be obtained in general, by indirect methods.

Four main methods of study have been used:—

(*a*) Examination of gut contents.

(*b*) Injection of dyes, drugs or labelled nutritional compounds into the host and the subsequent examination of the parasite for these substances.

(*c*) Histological and histochemical examination of a worm fixed *in situ*, to ascertain in particular,
(*i*) whether host tissues are eroded or otherwise affected by the presence of the parasite; and (*ii*) whether specific identifiable substances or cells, of host origin, such as haemoglobin, blood corpuscles or goblet cells, are present.

(*d*) Maintenance of worms *in vitro* and direct observation of them.

Of these methods, the last is likely to prove the most satisfactory but, unfortunately, culture methods for trematodes present special difficulties and not many data have yet been obtained by this means (see Chapter 9). The alternative methods must be used with caution, for abnormal results may sometimes be obtained if worms are allowed to come in contact with and ingest blood or tissue released from the host during the autopsy.[68]

The Alimentary Canal as a Biotope

Since a high proportion of trematode species are parasites of the intestine, the physiology of the alimentary canal is briefly dealt with here.

Until recently, it was believed that polysaccharides were broken down to monosaccharides, proteins to amino acids and fats to glycerol and fatty acids within the gut. It was also considered that only the relatively small molecules, produced as a result of these processes, were absorbed by the gut and that the quantity and proportions of these available at any one time was related to the feeding habits of the host.

Many of these concepts have been challenged and the evidence[289, 359] now suggests:—

(a) that disaccharides can be absorbed by the intestinal mucosa without first being hydrolysed to monosaccharides and that, following absorption, hydrolysis can subsequently take place in the epithelial cells;

(b) that products of protein digestion larger than amino acids can be absorbed;

(c) that the molar ratios of the pool of amino acids in the intestine remain relatively constant independent of the quality of the protein ingested in the host diet;

(d) that this equilibrium condition of the intestine, relative to the molar ratios of amino acids, is related to the release of amino acids from both shed and intact mucosal cells, so that a two-way traffic in amino acids can take place through the mucosa;

(e) that the mechanism of transport of amino acids and monosaccharides is not clear, but that it may be carried out by means of ' carriers ' whose nature is not understood;

(f) that the shedding and replacement of mucosal cells is a remarkably rapid process, the ' turnover time ' (i.e. the time taken for replacement of the number of cells equal to that in the total population) in all mammals studied—man included—being less than three days (table 1). From this it follows that at any time the intestinal contents must contain a significant proportion of material derived from degenerating cells of the intestinal mucosa as well as from digested host food.

The full implication of these findings concerning the physiology of intestinal parasites—and trematodes in particular—is not yet evident. Clearly they will be of particular significance to *in vitro* work (Chapter 9), and in considering the digestion and metabolism of trematodes.

TABLE 1

Turnover time of intestinal epithelium[359]

Area	Species	Turnover time (days)
Duodenum	rat	1·6
	cat	2·3
	man	1·8
Jejunum	rat	1·4
	mouse	1·8
Ileum	rat	1·4
	cat	2·8

Feeding

General mechanisms

Very little work has been carried out on feeding and digestion in trematodes.[149]

The presence of a well-developed oral sucker and pharynx in most species indicates that trematodes possess a powerful mechanism for sucking their food and forcing it into the caeca. In the case of intestinal trematodes, the semi-fluid food is probably drawn in by simple suction. Trematodes in tissue sites, feeding on blood and tissue, must rupture the tissues of the host to obtain food, and plugs of tissue are frequently drawn into the pharynx. In some strigeids (e.g. *Diplostomum phoxini*), the small pseudosuckers on either side of the oral sucker are used for attachment, with the result that the oral region is freed for browsing over the feeding surface. The efficiency of the feeding system in this parasite may be demonstrated by placing metacercaria in a viscous medium such as a yolk-albumen mixture (p. 156); within thirty minutes or so the gut may be seen to have become filled with food material.[9]

Nature of Food Materials

In general, the food material is likely to consist of what is most immediately available, such as intestinal contents, bile, blood or mucus. The picture is complicated, however, by the fact that (a) before reaching the site of definitive development many species (e.g. schistosomes) migrate through various organs so that the available food will vary, and (b) in certain sites, such as the bile ducts or intestine, erosion of host tissue may occur and cells and blood could also be part of the diet.

In the case of *Schistosoma mansoni*, for example, the adult lives in the blood stream and feeds solely on blood, as is evidenced by the presence of haematin in the gut.[263] The schistosomulae, on the other hand, pass through the lungs, heart and liver before reaching the definitive site in the mesenteric veins. It is especially interesting to note that the schistosomulae undergo virtually no growth (fig. 48) until they reach the liver, when a rapid increase in size and weight takes place. This suggests that the liver may provide particular food materials necessary for stimulation of growth—materials which are missing in the lungs and heart. Similarly, the metacercaria of *Fasciola hepatica* migrates through the intestinal wall and the liver, feeding as it goes, before reaching a bile duct. The young fluke apparently makes its way through the liver by using its oral sucker to bite deeply into a mass of hepatic cells by a ' pincer-like ' action.[70] Some blood is naturally freed during this process so that the food of the young *Fasciola* is probably a homogenate of liver cells and invading leucocytes, together with some erythrocytes and plasma.

The nature of the food of the adult *Fasciola hepatica*, when established in a bile duct, has been a matter of controversy.[72] Studies with sheep, in which the blood has been tagged with chromium radioisotope, ^{51}Cr, clearly show that some blood is taken into the gut of the fluke,[240] specimens in the larger ducts taking in as much as 0·58 ml. within two hours. Excess iron is reported to be eliminated through the excretory system and the tegument by virtue of forming a compound (ferric hydroxide—ferrous ascorbate) with Vitamin C, which enables it to be transferred to these sites.[239] The adult fluke contains substantial quantities of iron (0·5 per cent.—table 6).[127]

Histological studies further indicate that moving flukes abrade the superficial epithelium of the hypertrophied bile ducts with their spines; some workers hold[71] that the parasite feeds substantially on the homogenized tissue débris produced, rather than on blood.

These pieces of evidence are not necessarily in conflict, for it is likely that the food is a mixture of abraded tissue together with some freed blood. This is probably the picture, too, in some other liver parasites such as *Clonorchis, Opisthorchis* and *Dicrocoelium.* Most intestinal trematodes probably feed on semi-digested intestinal food contents, together with mucus and mucosal cells derived from the intestine and released blood.[127] Some flukes such as *Haplometra cylindracea* (lungs), *Haematoloechus medioplexus* (lungs), *Diplodiscus subclavatus* (rectum) in the frog appear to feed exclusively on blood[127] as, of course, also do schistosomes. In many sanguivorous trematodes the presence of degeneration products of haemoglobin, notably haematin, has been reported.[127, 263] Some species, such as *Paramphistomum microbothrium* (rumen),[75] can severely irritate the intestinal wall and in some cases even penetrate to the muscularis mucosa. In strigeids, the holdfast plays a major rôle in digestion by secreting histolytic enzymes and reducing the host tissues to a granular form suitable for digestion.[89] Thus, in *Cyathocotyle bushiensis*, a parasite in the rectum of ducks, the holdfast (which is rich in enzymes) has a marked effect on the mucosa. It frees the columnar epithelium from the underlying tissues in the vicinity of the parasite, and reduces its cells to a granular mush. This granular material is taken into the oral sucker and digested intracellularly within the parasite gut. In species other than strigeids both extracellular and intracellular digestion probably occurs, but there is little precise information on this point. There is some evidence (p. 16) that the intestinal cells may go through cycles of secretion and absorption.

Although substantial quantities of nutritive material must be obtained from digestion of semi-solid food in the gut, it has been shown experimentally (p. 13) that absorption of some nutrients can take place via the general body surface; in strigeids the tegument covering the holdfast has a microvillous structure and is almost certainly absorptive in nature.[88]

To this evidence must be added the fact that the tegument of several species (e.g. *S. mansoni*; table 2) is rich in acid and alkaline phosphatase.[233, 262] While the exact significance of phosphatases in cellular metabolism is not yet clearly understood, a relationship between the active transport of hexoses and the localization of these enzymes has been postulated.

Digestion

The gut cells of a trematode generally show high enzyme activity and also possess a high concentration of RNA, indicating

TABLE 2

Histochemical localization of acid and alkaline phosphatases in male and female Schistosoma mansoni[233]

o =absent;   + =slight activity;   + + + =high activity;   − =not relevant

Tissue	Acid phosphatase Male	Acid phosphatase Female	Alkaline phosphatase Male	Alkaline phosphatase Female
'Cuticle': dorsal	+ +	+ +	+ + +	+ + +
ventral	o	+ +	o	+ + +
Muscle	o	−	o	−
Parenchyma	+	+	+	+
Vitelline glands	−	+ +	−	o
Ovaries	−	+ + +	−	o
Testes	+ + +	−	o	−
Gut epithelium	+ + +	+ + +	o	o

that protein synthesis is occurring. The trematode caecum is thus an active site of enzyme formation and secretion and of digestion and absorption. Biochemical and histochemical techniques have been used to identify enzymes in a limited number of species (tables 2, 3). The following enzymes have been detected: proteases, a dipeptidase, an aminopeptidase, lipases, acid and alkaline phosphatases and esterases. Few detailed studies on the properties of these enzymes have been made. Tests for alkaline phosphatase were positive only when adenosine triphosphate or fructose-1·6-diphosphate were used as substrates.[273]

The enzyme system has been carefully studied in the case of *S. mansoni*.[321] This species, which occurs in the mesenteric veins of mammals, can degrade blood proteins. Presumably, on entering the gut of a schistosome the red cells are lysed either by a released acid or a lysin, or by unfavourable osmotic conditions. The globin so released is thus available for breaking down to peptides or amino acids. Homogenates and cell-free extracts of *S. mansoni* have been shown[322] to contain a protease able to release amino

TABLE 3

Nature of food and enzymes present in some species of Digenea
+ =present; o =absent; . =not tested for

				Protease	Lipase	Alkaline phosphatase	Acid phosphatase	Esterase	Amino-peptidase	
Species	Host	Location	Food							Reference
Haplometra cylindracea	frog	lungs	blood	+	.	+	+	+	.	127
Haematoloechus medioplexus	frog	lungs	blood	+	.	+	+	.	.	127
Diplodiscus subclavatus	frog	rectum	blood	.	.	+	+	.	.	127
Opisthioglyphe ranae	frog	intestine	tissue, mucus, blood.	.	.	+	+	+	.	127
Gorgoderina vitelliloba	frog	bladder	tissue, blood.	.	.	+	+	+	.	127
Fasciola hepatica	sheep	bile duct	tissue, blood	+	+	+	+	+	.	127 344
Schistosoma mansoni	mouse	venous system	blood	+	.	+	+	.	.	127 322
Cyathocotyle bushiensis	duck	rectum	tissue	.	.	+	.	+	+	89

nitrogen from globin or haemoglobin. Moreover, this protease is highly haemoglobin- and globin-specific, shows a marked maximum at pH 3·9 and is destroyed by boiling. When schistosomes digest globin both free amino acids and peptides are released. The latter contain the three basic amino acids histidine, tryptophan and

arginine, for which the worms appear to have special requirements.[278] There is some evidence for the existence of a second
protease with a pH optimum of about 6·0. Enzyme preparations
from this species failed to break down whole serum, which may
indicate the true absence of the requisite enzymes; this result,
however, could also indicate the presence of protease inhibitors.
Further evidence of the importance of haemoglobin in schistosome
nutrition is provided by the fact that the addition of red-cells is
markedly beneficial to *in vitro* cultures (p. 152). Evidence regarding the existence of phosphatases in the gut of *S. mansoni* is conflicting; some workers have reported the presence of both acid
and alkaline phosphatases;[85] others failed (table 2) to detect the
latter.[233, 262]

Attention has also been drawn to the rôle of the holdfast
of strigeids in digestion (p. 25). This organ has been shown to
secrete a complex of enzymes consisting of alkaline phosphatase,
several types of esterases and leucine aminopeptidase; it would
appear to provide interesting material for further study.[89]

From this brief survey it is clear that the nature of the food and
of the digestive enzymes of very few trematode species are known.
This is a field which particularly calls for further study.

4 : The Adult—General Metabolism

Respiration

General considerations

Oxygen consumption has been studied in various species and it has been shown that at all stages these trematodes utilize oxygen when it is available. Since miracidia and cercariae require oxygen in order to survive, they are obligatory aerobes. In adult trematodes the position is not clear, for although they are generally considered to be facultative anaerobes, there are few precise experimental data to support this view. A general review of respiration in trematodes has been made by Vernberg.[336]

It is not easy to obtain an accurate assessment of the oxygen requirement of an adult trematode *in vivo*. The most that can be done in experimental work is to prepare *in vitro* conditions as near as possible to those believed to exist *in vivo* and then to make the necessary measurements. These conditions are particularly difficult to assess in the case of the intestinal parasites for, although the oxygen tension of the intestinal lumen is known to be low for several hosts[265, 290, 344] in some cases approaching zero, that between the villi or in the crypts of Lieberkühn may be substantially higher. The thin layer of liquid in contact with the mucosa of the rat intestine has been shown[264] to have three times the oxygen tension of the bulk of the intestinal content. In the case of trematodes which live in bile (e.g. *Fasciola hepatica*), in the blood stream (e.g. *Schistosoma mansoni*), or in the molluscan kidney (e.g. *Proctoeces subtenuis*), the *in vivo* conditions can be reasonably assessed and reproduced experimentally. Apparatus suitable for routine measurements of the respiratory metabolism of parasites are the usual Warburg manometer, the polarograph using the oxygen electrode and the Scholander respirometer, or suitable modifications of these. For single, small specimens, the Cartesian diver has proved to be a valuable tool.[336]

Oxygen consumption

Quantitative results. Most of the factors which have been found to affect the rate of respiration in free living organisms affect oxygen consumption in trematodes. The most important of these are temperature, body size and oxygen tension. Figures for the oxygen consumption of adult trematodes are given in table 4.

TABLE 4

O_2 consumption of trematodes in atmospheric air

Species	Location	Q_{O_2} (μl.O_2/mg.) (dry wt./hr.)	Temp. °C.	Ref.
Fasciola hepatica	bile ducts	1·76–2·11	37·5	331
Fasciola gigantica	bile ducts	0·037	37·5	114
Schistosoma mansoni	mesenteric veins	3·05–10·15	37·5	23
Paramphistomum cervi	reticulum	0·03	38	172
Paramphistomum explanatum	bile ducts	0·84–1·34	37·5	112
Paragonimus westermani	lungs	0·74–0·86	37·5	253
Gynaecotyla adunca	intestine	0·032–0·022 *	30	338
Gastrothylax crumenifer	reticulum	0·30–0·33	37·5	112
Proctoeces subtenuis	organ of Bojanus (in mollusc)	1·68	15–17	98
Cotylophoron indicum	reticulum	0·023	37·5	114

* μl.O_2/hr./μgN.

Those for *P. subtenuis* can probably be taken as a reliable indication of the consumption *in vivo*; for the natural medium in this instance is isotonic with sea water, so that the latter can be conveniently used for experimental determinations. Some of the other figures may not be so reliable a guide to the *in vivo* respiration conditions. From these figures it can be seen that the oxygen consumption for *S. mansoni*, in the mesenteric veins of mammals, is many times higher than that of other species studied; it is further increased for this species when glucose is present in the incubation medium.

Influence of temperature. The effect of temperature on respiration is of special interest in poikilothermic animals, as there is evidence that the respiratory processes have become adapted to a particular range of temperatures.[334, 335, 340] Once the temperature is raised a few degrees above the temperature to which the parasite is normally subjected in its natural habitat, the respiration rate is depressed and the parasite dies. Thus the trematode *Gynaecotyla adunca*, from a bird (temp. 40° C.), showed increasing respiration rate up to 41° C., but could survive for only several hours at 45° C. (fig. 7). A turtle parasite, *Pleurogonius malaclemys*, showed an increased Q_{O_2} with increasing temperature up to 36° C., although at this temperature the rate declined after about ninety minutes. In contrast to these species, a fish parasite, *Saccocoelium beauforti* (natural temperature max., 30° C.), showed an increased respiration rate only up to 30° C. (fig. 7) and died within an hour at 41° C.[340] The temperature coefficients, expressed in terms of Q_{10}, also reveal the extent of the adaptation of trematodes to the temperature of their hosts; it has been shown experimentally in the case of free living animals that the Q_{10} values are lowest in the range of temperature which approaches that of the ' natural ' environment. Thus, at the temperature range normally encountered in the definitive host, the Q_{10} of trematodes tends to be low. In *P. malaclemys* (whose host has a wide temperature range) the Q_{10} was high only in the range 6°–12° C.; in *S. beauforti* the Q_{10} values were high up to 18° C., and in *G. adunca* up to 30° C.

Influence of size. The relationship between size and respiration has been studied for only a few species.[112, 338] In general, the respiration rate decreases during the life cycle as the worms grow larger. Thus, in the case of *Gynaecotyla adunca*, the body nitrogen increases to about 60 times that of the cercaria before the metacercaria encysts, and yet the oxygen consumption decreases by about 40 per cent.[338] In adult *G. adunca*, there is significant correlation between body nitrogen and respiration rate; a fourfold increase in size was correlated with a decrease of Q_{O_2} of about 40 per cent.

Influence of oxygen tension. Animals fall into two groups with regard to their response to varying oxygen tensions.[336] In the first group the rate of oxygen consumption depends on the oxygen

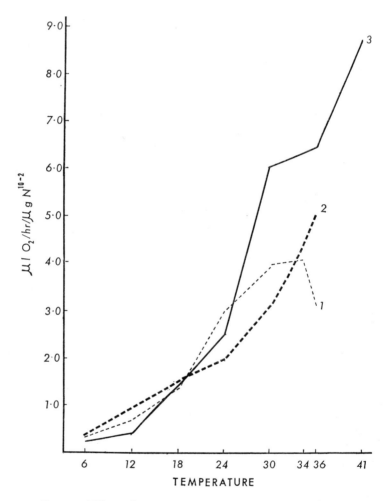

FIG. 7. Effect of temperature on oxygen consumption of trematodes from fish, reptiles and bird hosts. 1. *Saccocoelium beauforti* (from mullet); 2. *Pleurogonius malaclemys* (from marine turtles); 3. *Gynaecotyla adunca* (from shore birds). (after Vernberg and Hunter, 1961)

tension of the external environment and hence varies with the amount of oxygen available. Animals in this group have been termed ' conformers ', and they are able, in some way, to adjust their metabolic rate to the amount of oxygen available. The second group of animals includes the ' regulators ', i.e. those which respire at a rate which is independent of the external oxygen tension down to some critical level (known as the *critical oxygen level*, or Pc). Below this level the oxygen consumption declines rapidly.

Few investigations have been made into the effect of oxygen tension on respiration in adult trematodes. In the case of *Gynae-cotyla adunca* the organism is a conformer at levels of $1 \cdot 5$–$5 \cdot 0$ per cent. O_2, but below $1 \cdot 5$ per cent. O_2 becomes a regulator and maintains the same respiration rate in $0 \cdot 5$ per cent. O_2 as in $1 \cdot 5$ per cent. O_2.[336] Adults of *G. adunca* are unusual in being intestinal parasites of fish, turtles and shore birds. The levels $0 \cdot 5$–$1 \cdot 5$ per cent. may represent the range of oxygen tensions in this range of intestinal environments.

Significance of oxygen consumption. What is the significance of the figures for O_2 consumption? This question is not easy to answer but has been studied in several instances with the aid of cyanine dyes. These dyes inhibit the oxygen uptake of schistosomes both *in vitro* and *in vivo* (by injection of the hosts). Respiration of *S. mansoni* is almost completely inhibited (80 per cent.) by these dyes, but *the rate of glycolysis is unaffected* and the production of lactic acid (p. 39) is the same in the presence or absence of oxygen or of the cyanine dyes.[22] These results clearly indicate a largely anaerobic metabolism—a somewhat surprising result for an organism which lives in an environment with a relatively high oxygen tension. Perhaps *S. mansoni* has so adapted its metabolism in response to living in another anaerobic host site—such as the body cavity—occupied during an earlier phase of its evolution, but has retained from its free-living ancestors some ability to use oxygen. It may be that the organism still requires some oxygen to produce some essential metabolite. A somewhat similar situation occurs in *Fasciola* in which the metabolism is primarily anaerobic.[22] In both these species, the egg-shells are hardened by quinone-tanning (p. 62), a process involving oxygen, and this may account for some of the oxygen consumption in these cases.

Sufficient critical work has not been carried out to determine whether other trematodes follow a similar pattern.

Work with other parasites, particularly nematodes, suggests that dependence on respiratory metabolism in trematodes will be found to vary greatly from species to species. The adequacy of simple diffusion for passing oxygen to the central region of an animal can be calculated by the formula

$$Y = y_0 - \frac{a}{4k} (r_0^2 - r^2)$$

where $Y = pO_2$ (in atmospheres) at a distance r (cm.) from the centre (of a cylinder) and

$y_0 = pO_2$ (in atmospheres) at surface of animal
$r_0 =$ radius of cylinder (cm.)
$a = O_2$ uptake in ml./gm. wet wt./hr.
$k =$ diffusion constant for oxygen.

This formula has been applied to *Proctoeces subtenuis*[98] and in this small trematode (radius$=$0·25 mm.) an external oxygen pressure of about 24 mm./Hg would be sufficient to permit oxygen to reach the central tissues. Since the oxygen tension in the host is about 160 mm./Hg, it is clear that this species potentially can have an aerobic metabolism without necessarily requiring an oxygen transporting system. The position in other species appears not to be known.

Occurrence of respiratory pigments. Respiratory pigments with absorption spectra very close to those of mammalian haemoglobin have been described in eight species of trematodes; their spectra are given in table 5.

Chemical evidence such as is provided by (*a*) deoxygenation under reduced pressure or by the addition of sodium dithionite, (*b*) the formation of pyridine haemochromogen and (*c*) the fact that the pigment turns cherry red on exposure to carbon monoxide, confirms that these pigments are haemoglobin.

There is no evidence of the functioning of haemoglobin in trematodes. In *P. subtenuis* the haemoglobin seems refractory to deoxygenation and appears not to have a respiratory function; further investigation will be necessary to elucidate this question.

Most early workers have tended to compare haemoglobins in parasitic worms with mammalian haemoglobin (i.e. a circulatory haemoglobin with a sigmoid deoxygenation curve) and not with myoglobin, which in fact they more clearly resemble. It may be, then, that the haemoglobin in trematodes acts not only as an oxygen store but also as a transfer mechanism within and between cells.

Terminal oxidases. There is very little information on the terminal oxidase system in trematodes. A cytochrome oxidase system has been reported in *Fasciola*,[331] and *Schistosoma mansoni*

TABLE 5

Absorption maxima (in mμ) of trematode haemoglobins

Species	Oxyhaemoglobin	Reduced haemoglobin			Ref.
Telorchis robustus ⎫ Alassostoma magnum ⎭	575·0		—		350
Dicrocoelium lanceatum	?		—		331
Fasciola hepatica	580–581 543		—		331
Fasciola gigantica	586 540		—		116
Gastrothylax crumenifer	573·5 539·5		560		113
Cotylophoron indicum	585·2 555·8		—		116
Proctoeces subtenuis	579 543	418	558 431		99
Various mammals	576–578 540–542	412–415	555 430		99

contains a trace of cytochrome oxidase.[26] Cyanide, which normally depresses respiration in a cytochrome system, has been found to stimulate the respiration rate of *Paramphistomum cervi*;[172] in this case it may act by removing an inhibitor. These results may indicate that in some trematodes respiratory catalysts other than the cytochrome system are involved. Phenolases (o-diphenol:O_2 oxidoreductases) have been found in a number of trematodes where they are chiefly, if not entirely, concerned with the oxidation of phenolic compounds to form the quinones used in the hardening of the egg-shell (p. 62).

Carbohydrate Metabolism

Carbohydrate analysis and localization

Analysis. Like most other parasitic helminths, trematodes have a pronounced carbohydrate metabolism. This is reflected in the fact that they contain substantial quantities of stored carbohydrate (table 6), make extensive use of endogenous carbohydrate, have a high rate of transport of exogenous glucose into the tissues and a high rate of production of fatty acid *in vitro*. For example, *S. mansoni*, metabolizes an amount of glucose equivalent to one-fifth

TABLE 6

*Chemical composition of trematodes**

Species	Host	Location	dry matter as % fresh wt.	Chemical composition % dry wt.: inorganic ash	carbo-hydrate (glycogen)	protein	lipid	iron	Ref.
Haplometra cylindracea	frog	lungs	16·7	5·36	9·5	63	–	0·25	127
Haematoloechus medioplexus	frog	lungs	17·5	6·1	9·0	66	–	0·2	127
Diplodiscus subclavatus	frog	rectum	16·5	4·0	15·5	62	–	0·13	127
Gorgoderina vitelliloba	frog	bladder	19·3	4·5	8·7	–	–	0·2	127
Fasciola hepatica	sheep	bile ducts	19·5	5·5	16·0	59	12–13	0·5	127, 344
Fasciola gigantica	buffalo	bile ducts	22·4	–	25·4	66	12·85	–	110, 115, 111
Paramphistomum explanatum	buffalo	bile ducts	25·7	–	25·0	53	4·51	–	109, 110, 111
Gastrothylax crumenifer	buffalo	reticulum	27·9	–	30·3	49	1·36	–	109, 110, 111
Schistosoma mansoni (male)	mouse	venous system	–	–	14·29	50	–	–	344
Schistosoma mansoni (female)	mouse	venous system	–	–	3·5	65	–	–	344
Gynaecotyla adunca	bird	intestine	–	–	1·6–7 (wet wt.)	–	–	–	337

* Factors such as age, degree of maturity and nutritional condition of host must be known for these figures to be of any real significance.

of its dried weight within the period of one hour.[23] *Paramphistomum explanatum* uses 36 per cent. and *Gastrothylax crumenifer* 42 per cent. of its glycogen content in ten hours.[109] Under starvation conditions, *F. hepatica* uses up 20 per cent. of its glycogen within five hours, and *F. gigantica* 35 per cent. within ten hours.[115] The chemical composition and metabolism of only a few species have been studied in detail; these are listed in table 6. Figures for chemical analysis of parasites must always be accepted with caution; for the carbohydrate composition may fluctuate markedly with the nutritional condition of the host (i.e. according to whether or not the latter has been feeding just before autopsy).

The main polysaccharide found in trematodes is glycogen, which does not appear to differ significantly in its properties from glycogens in vertebrates, although there is some evidence of polydispersity characterized by two regions of sedimentation coefficients.[346]

Polysaccharide localization. The localization of glycogen, as demonstrated histochemically, is very uniform in the species

TABLE 7

Polysaccharide utilization and synthesis by Fasciola hepatica
(*after Mansour*[211])

Experiment No.	Initial	Glycogen content (μ moles glucose/gm. wet wt.)	
		After starvation overnight	After feeding for 24 hr.
1	275	135	238
2	151	51	187
3	204	123	204
4	160	76	165
5	212	58	153

studied.[6, 20, 346] The most important region of storage appears to be the parenchyma, where the cells are filled with granules of varying size; this region appears to serve as a carbohydrate reserve analogous to that of the liver of vertebrates. Quantities of glycogen also occur in the muscular organs, especially the suckers and the cirrus pouch. The glycogen is deposited in the non-contractile parts of the muscle cells, especially the parenchymal sheath, but not in the contractile part. Ovarian eggs also contain some glycogen, and the mature vitelline cells somewhat more.

After twelve hours' starvation some loss of glycogen can be detected histochemically, but after twenty-four hours the parenchyma glycogen disappears almost entirely.[346] In worms cultured *in vitro* in glucose and serum a rapid synthesis of glycogen occurs, which can be detected histochemically as well as chemically (table 7).

Carbohydrate metabolism: General

Little is known concerning many important details of trematode metabolism. In several species it has been demonstrated that adult trematodes can make use of hexoses such as glucose and mannose, but not ribose.[211, 341, 346] Under starvation conditions, endogenous glycogen is used up but, as has been mentioned, can be resynthesized rapidly when exogenous food materials become available (table 7).

The glucose intake in *F. hepatica* has been shown to be the same in worms in which the gut was ligatured as in intact worms,[211] so that clearly glucose can be taken in through the tegument. In this connection it is interesting to note the distribution of phosphatases in trematodes.

Two types of phosphatases are known. The ' acid ' phosphatases are a group of enzymes found in plant and animal tissues which act optimally on monoesters of phosphoric acid at pH values near 5; an especially rich source is the human prostate gland from which the enzyme has been extracted in a pure form. ' Alkaline ' phosphatases, on the other hand, act optimally at pH values near 9; and most known alkaline phosphatases require the presence of divalent cations, such as Mg^{++} for enzymic activity.

In vertebrates, alkaline phosphatases are present at high concentrations in sites such as the intestinal mucosa and proximal renal tubules, where active transport of glucose is known to occur and the existence of a functional relationship has been postulated. In *S. mansoni* acid and alkaline phosphatases have been detected in the tegument but alkaline phosphatase is absent from the gut (table 2). If the functional relationship between alkaline phosphatase and glucose transport is confirmed, this will provide further evidence that in this species glucose transport takes place mainly through the tegument. In *Clonorchis sinensis* (in the liver), on the other hand, a high level of acid phosphatase is found in most tissues of the body including the tegument, and alkaline phosphatase activity is low.[201] In this species the phosphatases have been shown to consist of several isoenzymes which have different pH optima and can be separated electrophoretically.

Trematodes appear to have a predominantly fermentative type

of metabolism. This conclusion is based on the fact that the rate of carbohydrate consumption and the production of metabolic products in the form of fatty acids and CO_2 are almost identical under anaerobic and aerobic conditions. As was earlier pointed out (p. 33), this is perhaps understandable in the case of *F. hepatica*, which lives in the essentially anaerobic environment of the bile duct; but is difficult to account for in the case of *S. mansoni* which lives in the mesenteric veins—an essentially aerobic environment.

End-products of carbohydrate metabolism

In predominantly aerobic organisms, such as vertebrates, the end-product of glycolysis (table 8) is almost exclusively lactic acid (from pyruvic acid). This is produced as a result of muscle contractions so rapid that they are essentially carried out under anaerobic conditions.

This anaerobic phase is followed normally by an aerobic phase in which much of the pyruvic acid undergoes oxidative decarboxylation to acetyl coenzyme A and is used further in the Krebs cycle (table 10) to CO_2 and H_2O. If aerobic conditions do not become available immediately after the anaerobic phase, an ' oxygen debt ' is said to accumulate.

It has been found that in parasitic organisms the end-products of the anaerobic metabolism show some degree of variation. In trematodes, however, the range of end-products produced shows an unusually high degree of variation. Thus, the major end-product of the carbohydrate metabolism of *S. mansoni* is DL lactic acid, 81–91 per cent. of the glucose used being converted into this substance.[23] In *F. hepatica*, on the other hand, lactic acid forms only 4–9 per cent. of the metabolized carbohydrate, the remaining 81–95 per cent. being the volatile fatty acids, propionic and acetic acids, in a ratio of 3 : 1.[211] In *Clonorchis sinensis* 50 per cent. of the acid produced is lactic acid.[251]

In contrast with vertebrate tissues, where the lactic acid is further metabolized when oxygen becomes available, the fatty acids of trematodes are almost entirely excreted without further metabolizing. Thus there is virtually no ' oxygen debt ' in trematodes—at least in those species so far studied.

TABLE 8

Embden-Meyerhof pathway of anaerobic glycolysis*

1. Glycogen$+nH_3PO_4$ $\xrightarrow{\text{α-glucan phosphorylase}}$ n glucose 1-phosphate

2. Glucose 1-phosphate$+$glucose 1, 6-diphosphate $\xrightarrow{\text{Phosphoglucomutase}}$ glucose 1, 6-diphosphate$+$glucose 6-phosphate

Alternatively, starting with glucose, 1 and 2 are replaced by 3.

3. Glucose$+$ATP $\xrightarrow{\text{Hexokinase}}$ glucose 6-phosphate$+$ADP

4. Glucose 6-phosphate $\xrightarrow{\text{Glucosephosphate isomerase}}$ fructose 6-phosphate

5. Fructose 6-phosphate$+$ATP $\xrightarrow{\text{Phosphofructokinase}}$ fructose 1, 6-diphosphate $+$ADP

6. Fructose, 1, 6-diphosphate $\xrightarrow{\text{Aldolase}}$ dihydroxyacetone phosphate$+$D-glyceraldehyde 3-phosphate

7. Dihydroxyacetone phosphate $\xrightarrow{\text{Triosephosphate isomerase}}$ D-glyceraldehyde 3-phosphate

8. D-glyceraldehyde 3-phosphate$+$dehydrogenase$+$NAD $\xrightarrow{\text{Glyceraldehydephosphate}}$ 3-phospho-D-glyceric acid-enzyme complex dehydrogenase $+NADH_2$

9. 3-phospho-D-glyceric acid-enzyme complex$+H_3PO_4$ $\xrightarrow{\text{Glyceraldehydephosphate}}$ 1, 3-diphospho-D-glycerate$+$dehydrogenase dehydrogenase

10. 1, 3-diphospho-D-glycerate$+$ADP $\xrightarrow{\text{Phosphoglycerate kinase}}$ 3-phospho-D-glycerate$+$ATP

11. 3-phospho-D-glycerate$+$2, 3-diphospho-D-glycerate $\xrightarrow{\text{Phosphoglyceromutase}}$ 2, 3-diphospho-D-glycerate$+$2-phospho-D-glycerate

12. 2-phospho-D-glycerate $\xrightarrow{\text{Phosphopyruvate hydratase}}$ phosphoenol-pyruvate $+H_2O$

13. Phosphoenol-pyruvate$+$ADP $\xrightarrow{\text{Pyruvate kinase}}$ pyruvate$+$ATP

If O_2 is absent, the $NADH_2$ from Reaction 8 is oxidized in muscle by Reaction 14 and the end-product is lactate:

14. Pyruvate $\xrightarrow{\text{Lactate dehydrogenase}}$ L-lactate

Sum of reactions 3–14: glucose$+$2 ADP$+$2H_3PO_4 $=$2 lactic acid$+$2 ATP $+$2H_2O

* Enzyme terminology as recommended by the Enzyme Commission.[80]

*Intermediary metabolism**

(a) *Embden-Meyerhof pathway*. What few data are available suggest that the primary catabolism of carbohydrate in trematodes follows a scheme of phosphorylating glycolysis comparable to that of vertebrate tissue and known as the Embden-Meyerhof pathway. This is based on a series of steps, involving a succession of enzymes (table 8), the general pattern of which is well known. The evidence for the existence of the Embden-Meyerhof pathway consists essentially of demonstrations of the enzymatic steps and identification of intermediate compounds in the pathway. It must be pointed out, however, that the demonstration of a glycolytic enzyme in worm tissue does not *necessarily* mean that that enzyme is substantially involved in the metabolism, for there are examples from other organisms (such as nematodes) in which an enzyme (e.g. lactic dehydrogenase) has been detected which appears not to play any part in the metabolism.

A number of enzymes of the Embden-Meyerhof pathway have been detected mainly in *Schistosoma mansoni*[321]: several hexokinases, phosphoglucose isomerase, aldolase, phosphofructokinase, triosephosphate isomerase, phosphoglyceraldehyde dehydrogenase and lactic dehydrogenase. Some striking differences between comparable enzymes from parasite and host have been demonstrated using immunological techniques. Enzymes, in common with other proteins, can act as antigens (see Chapter 10) and possess immunological specificities. In addition, however, they possess specificity for a substrate or a group of substrates. It has been found that, although enzymes catalyzing the same biochemical reactions have similar substrate specificities, they may be different in other ways, such as optimum pH.[25] This result came as something of a surprise to biochemists, and such differences were first demonstrated for the glycolytic enzymes. The terms *isozyme* or *isoenzyme* have been proposed to describe the different molecular forms in which protein may exist with the same enzyme specificity.

---

* Throughout this text, in general, enzymes have been given the names used by the investigator concerned. Enzyme nomenclature has now been standardized by an Enzyme Commission and the student is referred to Dixon and Webb[80] for details of current usage. The Enzyme Commission nomenclature of the enzymes of the Embden-Meyerhof pathway and the Krebs cycle are given in tables 8 and 10.

Thus, when purified rabbit lactic dehydrogenase (LDH) was injected into birds, the antiserum obtained, although it inhibited the activity of the original antigen, did not inhibit LDH activity of *S. mansoni* and *S. japonicum*, and had a reduced inhibitory effect on the comparable enzyme from rat muscle.[214] If the LDH was preincubated with nicotinamide-adenine dinucleotide (NAD), it almost completely prevented inhibition by the antiserum—a result which suggested that the antibody reacts in the vicinity of the site of NAD-enzyme interaction.

It has also been found that certain trivalent antimony drugs[213] inhibit phosphofructokinase activity in *S. mansoni* and yet mammalian phosphofructokinase is relatively insensitive to these drugs. This is an encouraging result from the point of view of chemotherapy, because serious doubt has been expressed as to whether the enzymes of hosts and parasites would be sufficiently different in properties to allow the use of selective antiparasitic drugs.

Further indications that trematodes follow a glycolysis sequence similar to the Embden-Meyerhof pathway in vertebrates are provided by the identification of intermediary substances associated with glycolysis. Unfortunately, the number of investigations carried out has been small. Thus, by means of the elegant technique of using ^{14}C-labelled glucose, the following substances have been detected: hexose phosphate, phosphoenol-pyruvic acid, alanine (an amino acid derived from pyruvic acid by transamination) and lactic acid[21] (table 9).

(*b*) *Krebs* (*citric acid*) *cycle* (table 10). This cycle represents the principal metabolic pathway for the aerobic oxidation of pyruvic acid in mammalian tissues. It has been shown to occur in a number of invertebrates including some nematodes[117] and cestodes, although sometimes in a modified form. The significance of the cycle to an organism is, however, much greater than that indicated by the mere oxidation of pyruvic acid, for it is essentially the point at which the carbohydrate, fat and protein metabolisms can meet. Thus, since α-ketoglutaric, oxaloacetic and pyruvic acids can be formed easily from the amino acids glutamic and aspartic acids and alanine, the cycle is also concerned in the oxidation of some of the protein intermediates. Similarly, many other amino acids can be broken down to the substances which form the intermediate compounds of the cycle, such as fumarate, succinate, oxaloacetate and α-ketoglutarate.

Thus most of the carbon atoms of the chemical intermediates which form the chemical constituents of the cell pass through the Krebs cycle. The explanation for this is that the 4-carbon intermediates in particular are closely involved in other metabolic pathways, e.g. those concerned in the synthesis of the ribose material in the formation of nucleic acids. There is evidence for a complete Krebs cycle in some trematodes and for a partial cycle

TABLE 9

Percentage incorporation from ¹⁴*C-glucose in soluble intermediates of the Embden-Meyerhof sequence and Krebs cycle in* Fasciola hepatica
(After Bryant and Williams[21])

Intermediate	Adults	Miracidia	
Hexose phosphate and phosphoenol-pyruvate	15	21	associated with glycolytic activity
Alanine	10	26	
Lactate	18	30	
Succinate	17	0	associated with Krebs cycle activity
Fumarate	1	0	
Malate	4	0	
Citrate	4	0	
Glutamate	10	0	
γ-Aminobutyrate	6	0	
Dissacharide	3	7	

in others, but terminal oxidation is very deficient. Thus, when *F. hepatica* is incubated with ¹⁴C-succinate it becomes incorporated into the following intermediates: fumarate, malate, citrate, glutamate, γ-aminobutyrate and lactate, thus indicating an almost complete cycle.[21] Malonate, which competitively inhibits succinic dehydrogenase in the Krebs cycle, inhibits respiration in several species.[331, 339]

In *Gynaecotyla adunca*, on the other hand, evidence has been found for only a partial Krebs cycle. Thus, succinate, malate and oxaloacetate stimulated respiration in this species, whereas the intermediates citrate, isocitrate and pyruvate were without effect.

(*c*) *Effect of serotonin on carbohydrate metabolism.* Some interesting studies have been carried out with the substance serotonin (5-hydroxytryptamine) a known neuro-humor of smooth muscle. This substance, which has been found in minute quantities in

TABLE 10

The Citric acid or Krebs cycle*

1. Acetyl-CoA + oxaloacetate $\xrightarrow{\text{Citrate synthase}}$ citrate + CoA

2. Citrate $\xrightarrow{\text{Aconitate hydratase}}$ *cis*-aconitate + H_2O

3. *cis*-aconitate + H_2O $\xrightarrow{\text{Aconitate hydratase}}$ Ls-isocitrate

4. Ls-isocitrate + NADP $\xrightarrow{\text{Isocitrate dehydrogenase}}$ oxalosuccinate + NADPH$_2$

5. NADPH$_2$ + $\frac{1}{2}O_2$ $\xrightarrow[\text{and cytochrome oxidase}]{\text{NADPH}_2 \text{ cytochrome } c \text{ reductase}}$ NADP + H_2O

6. Oxalosuccinate $\xrightarrow{\text{Isocitrate dehydrogenase}}$ 2-oxoglutarate + CO_2

7. 2-oxoglutarate + oxidized lipoate $\xrightarrow[\text{pyrophosphate}]{\text{Oxoglutarate dehydrogenase with thiamine}}$ 6-S-succinyl-hydrolipoate

8. 6-S-succinyl-hydrolipoate $\xrightarrow{\text{Lipoate acetyltransferase}}$ succinyl-CoA + dihydrolipoate

9. Succinyl-CoA + H_2O $\xrightarrow{\text{Succinyl-CoA hydrolase}}$ succinate + CoA

10. Succinate + $\frac{1}{2}O_2$ $\xrightarrow{\text{Succinate dehydrogenase and cytochrome oxidase}}$ fumarate + H_2O

11. Fumarate + H_2O $\xrightarrow{\text{Fumarate hydratase}}$ L-malate

12. L-malate + NAD $\xrightarrow{\text{Malate dehydrogenase}}$ oxaloacetate + NADH$_2$

Sum of reactions: Acetyl-CoA + $2O_2$ = $2CO_2$ + H_2O + CoA

* Enzyme terminology as recommended by the Enzyme Commission.[80]

Fasciola, has been found to have a stimulating effect on the rhythmical movement of *Fasciola*.[207-212, 215, 216] The effect, which was found to be peripheral and not mediated through the central ganglia, occurred both aerobically and anaerobically.

Stimulation of parasite movement under anaerobic conditions by low concentrations of serotonin results in an increase in glucose uptake, glycogen breakdown and a 2- to 10-fold increase in lactic acid production, but no change in the production of volatile fatty acids. Serotonin may act by stimulating phosphofructokinase activity, since incubating serotonin-treated worms in fructose-6-phosphate causes a decline in the concentration of this substance and an increase in fructosediphosphate.[212] Alternatively, it may act by stimulating muscle contraction, a process for which fructose-6-phosphate could supply the energy.

Summary of carbohydrate metabolism

From the data given above, it is clear that carbohydrate is probably the most important energy source for trematodes. It is also evident that although a Krebs cycle may occur, there is, in general, only incomplete oxidation of energy-yielding substrates to fatty acids such as lactic, acetic and propionic acids, instead of oxidation to CO_2 and H_2O as in most free-living organisms.

As compared with free-living organisms, then, trematodes appear to have a wasteful and inefficient metabolism. It must be remembered, however, that, unlike the latter, they have access to almost unlimited food supplies, with the result that maintenance of a complex food-obtaining mechanism (which would require energy expenditure) is no longer necessary for survival. It follows from this that one form of economy which might be of advantage to the parasite would be to refine the catabolic system. We can ask, how much then of the normal Krebs cycle is essential for a trematode's existence? Sufficient energy is undoubtedly available from the Embden-Meyerhof pathway; but we can also raise the modified question, how many of the *synthetic* reactions of the Krebs cycle are *essential* for the synthesis of biological materials in the parasite and how many can be taken over by the host? Are any Krebs cycle intermediates available from the host and can the parasite use them in the form provided? Alternatively, can the host use the partially oxidized materials, such as fatty acids, released by the parasite? If this should prove to be possible, then a most interesting host-parasite relationship might be revealed; namely, that the food materials taken in by the parasite are used in the first instance anaerobically by the parasite and what is left is used aerobically by the host whose tissues have oxygen freely available. This speculation is probably an over-simplification of an extremely complex problem and further work may reveal an even more intricate metabolic relationship between host and parasite.

Protein Metabolism

General comments

So little precise exerimental work has been carried out on the protein metabolism that it is not possible to give more than a very general and tentative account of this aspect of trematode physiology. As discussed earlier (p. 22), our concept of the nature of certain host environments—especially the intestine—and the factors governing absorption of amino acids from them have changed radically within the last few years. A knowledge of the amino acid composition of these environments and the extent to which this composition fluctuates, together with information on the extent to which the tegument participates in amino acid uptake, is necessary before an accurate picture of protein intake can be composed. The rôle of the trematode intestine in protein digestion and absorption has long been accepted, of course, but the possibility that the tegument plays a major rôle presents a comparatively new concept in trematode physiology.

Both larval and adult trematodes possess remarkable powers of protein synthesis. A single miracidium of *S. mansoni*, for example, may produce more than 20,000 cercariae. Adult trematodes probably grow only slowly, once they become sexually mature, but the level of their protein synthesis is apparent from examination of species such as *Fasciolopsis buski*, the adults of which may produce up to 25,000 eggs daily.

Chemical composition

Comparatively little is known about the composition of the tissue proteins or the protein metabolism in trematodes. Data for the protein content are given in table 6. These figures are generally based on analysis of total nitrogen multiplied by a factor of 6·25. It must be stressed that this factor was derived from analyses of vertebrate tissues, and that the non-protein nitrogen may be sufficiently different in other groups for a substantial error to be introduced. The tabulated figures must, therefore, be accepted with caution. The nature of proteins in trematodes has been little studied, with the exception of the egg-shell which is dealt with separately (p. 59).

Protein intake

Adults invariably live in an environment that is relatively rich in protein together with some free amino acids. They can undoubtedly degrade and ingest complex proteins. Little is known, however, of the digestive enzymes involved in these processes, although there is some evidence of proteolytic enzymes in the gut. In blood flukes, such as *S. mansoni*, haematin has been detected spectroscopically in the gut[263] and there is no doubt that host blood cells can be digested. This fact is also evident from *in vitro* experiments (p. 152), in which the addition of red cells to tissue culture media greatly enhances the degree of maturation and the period of survival.[54]

It has been shown (p. 27) that schistosomes possess a protease with a high specificity which releases tyrosine from globin and haemoglobin; this enzyme has a pH of 3·9.[321] A lyophilized worm homogenate (i.e. a crude enzyme preparation) has also been found to degrade haemoglobin so that both amino acids and peptides are produced. Arginine, histidine and tryptophane are present in these peptides—an interesting observation in view of the fact that schistosomes have been shown to use these amino acids.[278]

That strigeid trematodes can digest complex proteins may be demonstrated experimentally with *Diplostomum phoxini*. Metacercariae of this species are found in the brain of the common minnow, *Phoxinus phoxinus* (p. 156). If removed aseptically and cultured in various liquid nutrient media, survival may be for five or six days but the total number of dividing cells is small and maturation is never attained. When, however, a semi-solid egg-yolk/albumen[9] or yeast/albumen[363] mixture is used, the gut fills rapidly, the mitosis rate increases suddenly to a level approaching that found in the bird host, and maturation is achieved, although only abnormal eggs are produced *in vitro*—a result suggesting some nutritional deficiency.

Amino acid requirements

It is difficult to obtain accurate information on the amino acid requirements of trematodes. Experiments must be carried out *in vitro* and, as pointed out in Chapter 9, a dilemma arises because satisfactory culture media can be developed only after the nutritional requirements are more precisely defined. Nevertheless, synthetic, protein-free culture media containing complex mixtures have been developed (p. 151) and the best of these have enabled *Schistosoma mansoni* to survive and achieve egg production for several weeks. The defects of this chemically defined medium were: (*a*) it did not permit indefinite survival *in vitro*, (*b*) after two or three weeks in it worms showed noticeable deterioration, (*c*) eggs produced were abnormal and sterile, failing to give rise to live miracidia.[277] These limitations must therefore be borne in mind when the results are being considered.

One interesting result obtained from *in vitro* experiments showed that the values of many amino acids were increased, not decreased (table 11) when specimens of *S. mansoni* were incubated in the culture media. The principal increase took place in the alanine, proline and ornithine levels, and the principal utilization *in vitro* occurred in the basic group (histidine, tryptophan, arginine) and the amino sugar, glucosamine. Several explanations of these results are possible. It could be argued that the worms were gradually deteriorating during culture and that necrobiosis releases amino acids from the trematode proteins. Alternatively, the parasites might have been excreting amino acids (p. 50). The ornithine could be produced from arginine, for urea is also produced and the highest values of ornithine in a particular culture were found to coincide with the lowest level of arginine.[277] This suggests that this trematode has a source of arginase, an enzyme usually found in high concentration in the liver cells of vertebrates. Accumulation of proline may be related to a proline-urea cycle but there is no experimental evidence to support this view.

Experiments with labelled ^{14}C show that the newly produced alanine had its origin in carbohydrate, and was eventually incorporated into the proteins of the worm; about 6 per cent. of the carbon in labelled glucose appeared in the new synthesized alanine.

Experiments such as this tend to assume that the absorption of amino acids is only a matter of active transport. Studies on amino

TABLE II

Increase or decrease of amino-acid content in a synthetic medium (SM − 1) or Hanks saline after incubating Schistosoma mansoni *in it for 48 hours* (Data from Senft[277])

Substance	SM-I medium	Hanks
Aspartic acid	trace	trace
Threonine	+	+
Serine	+ +	+ +
Proline	+ + + +	trace
Glycine	+	+
Alanine	+ + + +	+ + + +
Valine	+	+
Methionine	+	trace
Isoleucine	+	+
Leucine	+	+
Ornithine	+ + +	o
Lysine	o	o or trace
Histidine	decreased	o or trace
Tryptophan	decreased	o
Arginine	decreased	o
Urea	+ + +	o or trace

acid uptake by cells or tissues (such as the intestine or kidney) in other organisms have shown that the problem is much more complex than it was formerly thought to be; for under certain conditions amino acids in cells can ' leak out ' into a medium making contact with them. Thus, in the case of the mammalian intestinal mucosa, it has been shown that:

(a) L-acids are absorbed more rapidly than D-acids;

(b) there is a special amino acid transfer mechanism which can transfer amino acids against a concentration gradient;

(c) if more than one amino acid is present, there is competition for the transfer mechanism so that the presence of one amino acid may inhibit the uptake of another;

(d) the rate of absorption of an amino acid is not proportional to its concentration;

(e) the transfer of amino acids by the intestine may be inhibited by 2:4 dinitrophenol.

The above results suggest that some ' carrier ' mechanisms are involved, but the nature of the processes at work is unknown; for a general discussion the reviews of CHRISTENSEN,[49] WILBRANT and ROSENBERG[355] or SMYTH, D. H.[289] should be consulted. The carrier may prove to be a substance confined to a single locus in the cell, which can combine with the substance carried, thus facilitating its passage through this region. Again, it could be an enzyme fixed in a particular part of the cell which converts the substance into a form more easily transportable through the cell. Still another possibility is that it is an ' expansile ' carrier in the form of a protein molecule, capable of folding and unfolding and so transferring attached substances.

In the intestinal lumen (p. 22) the *molar ratio* of amino acids remains relatively constant, and it is clear that there is a two-way traffic of amino acid transport (i.e. amino acids can equally well ' leak ' *from* the mucosa into the lumen or be transported from the lumen in the opposite direction). Experiments on membrane transport in cestodes[132, 252] suggest that the same general principles apply and that if cestodes are placed in 'unsuitable ' amino acid mixtures a ' leak ' from the worm to the medium can occur in the same manner.

Since the trematode tegument (p. 11) shows some structural similarities to that of cestodes, and is permeable both to carbohydrate and to amino acids, it is probable that the two-way transport of amino acids across the tegument or gut cells of trematodes is governed by the same principles. This is supported by the fact that amino acids can pass from *Fasciola* into serum, when incubated in it, even when the mouth of the worm is tied.[171]

It is clear, then, that in the light of the new concepts of membrane transport, much more detailed and carefully controlled experiments must be carried out before reliable information on amino acid requirements of trematodes can be obtained. Some significant experiments on transamination have been carried out, as described below, but experiments on the competitive uptake of amino acids, comparable with those performed on cestodes, do not appear to have been carried out on trematodes.

Intermediary metabolism

(a) *Transamination*. The major studies on the amino acid metabolism in trematodes have been concerned with transaminating enzymes in the tissues. In order to understand these studies, it will be necessary to know something of the significance of the process known as ' transamination ', the mechanism of which has been worked out mainly on mammalian tissues.

It was found that if an amino acid (e.g. leucine) labelled with ^{15}N was administered to a mammal, a number of amino acids in the proteins of the liver contained ^{15}N. It can be concluded, therefore, that the organism can make use of the nitrogen of leucine for synthesis of other amino acids. This is an important concept in understanding the significance of ' essential ' and ' non-essential ' amino acids. Two general mechanisms are known for the utilization of the amino nitrogen from one amino acid for the formation of another amino acid. The first mechanism involves the separation of nitrogen by the process known as *deamination* and the use of the free ammonia so formed for the synthesis of other amino acids. The second mechanism is *transamination*, whereby the α-amino nitrogen of one amino acid is transferred directly to a keto-acid which acts as a skeleton or framework, so to speak, for the formation of another amino acid.

A number of transaminases have been demonstrated in *F. hepatica* and *Schistosoma japonicum*.[59, 135, 171] Thus, if *Fasciola* is incubated in serum containing known amounts of alanine and α-ketoglutaric acid, after six hours the amount of glutamic acid and pyruvic acid is increased four times and three times respectively (fig. 8). This provides clear evidence for the existence of powerful transaminases, and at the same time demonstrates in a convincing manner the permeability of the tegument to the substance concerned. In *S. japonicum*, glutamic-pyruvic and glutamic-oxaloacetic transaminases with *p*H optima in the range $7 \cdot 2 – 7 \cdot 5$ have been demonstrated.[135] It is highly significant that g-p-transaminase from *S. japonicum* was inhibited by antimony-containing drugs, which may explain the effectiveness of these in treating schistosomiasis. It is particularly interesting to note that these drugs have no effect on the comparable enzymes from mouse liver, indicating that, as has already been pointed out (p. 42), there may be differences between parasite and host enzyme.

It has further been demonstrated that the transaminase activity in *Fasciola* falls off rapidly when this organism is cultured in simple saline solutions, but not in serum, and that the activity can be almost restored by the addition of the prosthetic group pyridoxal phosphate (vitamin B_6), which acts as a co-factor for these enzymes, as it does in higher forms (fig. 9). This may

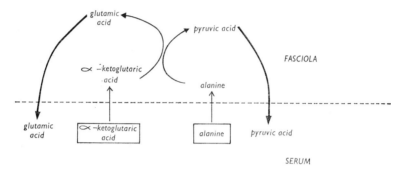

FIG. 8. Diagram of the uptake of alanine and α-ketoglutaric acid by *Fasciola hepatica* and their subsequent trans-amination. (after Kurelec and Ehrlich, 1963)

account for the value of this substance in certain trematode culture media (p. 157).

(*b*) *End-products of nitrogen metabolism.* Nitrogen appears to be excreted mainly in the form of ammonia and uric acid,[111] but urea has also been reported in one instance.[277] Figures for nitrogen excretion (table 12) are generally in keeping with the conclusion that small animals have a higher rate of metabolism than larger ones. Thus, *F. gigantica*, the largest of the parasites listed, excretes 2·5 per cent. of its total body nitrogen, whereas the smallest species, *P. explanatum*, excretes the highest amount, viz. 4 per cent. From the few details available it is clear that the catabolic nitrogen metabolism is very poorly understood.

It is generally held that ammonotelic organisms live in habitats where water to remove the end-products is freely available, whereas ureotelic organisms usually occur under conditions where they have to economize with water usage. Helminth parasites in

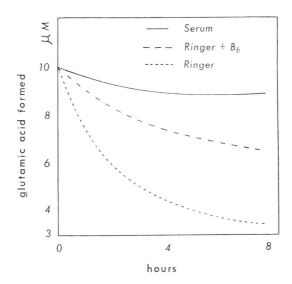

FIG. 9. Activity of transaminase from *Fasciola hepatica* after 8-hr. incubation of specimens in horse serum, Ringer and Ringer+Vitamin B_6 measured in μM of formed glutamic acid. (after Kurelec and Ehrlich, 1963)

TABLE 12

Percentages of total body nitrogen excreted in a period of 12-hour starvation under aerobic conditions at 37° C.
(After Goil[111])

Species	Ammonia	Uric Acid	Creatinine, Urea
Paramphistomum explanatum	4·0	0·077	Not detected
Fasciola gigantica	2·51	0·10	,,
Gastrothylax crumenifer	2·9	0·033	,,

general are primarily ammonotelic, and trematodes are apparently no exception.

(*c*) *Presence of histamine.* An unusual feature, found in the nitrogen analyses of trematodes is that histamine occurs in substantial quantities; high histidine decarboxylase activity is also reported. Histamine is an amine with irritant properties, which occurs commonly in fish and is present in most vertebrate tissues but not in free-living platyhelminths. Its presence in trematodes is difficult to explain.

Lipid Metabolism

Little is known regarding the lipid metabolism of trematodes. Analyses of lipid content in several species are given in table 6. In general, the content varies considerably. During starvation conditions the fat content tends to rise, sometimes substantially. This can be explained by the fact that higher fatty acids are by-products of the carbohydrate metabolism. Esterase activity has been demonstrated in several trematodes. Lipase has been detected in *Fasciola*, and a cholinesterase has been detected in the nervous system of a number of species (p. 20). A non-specific esterase has been demonstrated by histochemical means to be present in the forebody glands and adhesive organs of the meta-cercariae and adults of several species of strigeids.[89, 174]

It appears that nothing is known of the pathways of lipid metabolism.

Nutrition

Information on the nutritional requirements of trematodes is extremely meagre, and we have little detailed knowledge of the kind or quantity of protein, lipid or carbohydrate required. As might be expected, a low protein diet in a host results in a reduction in egg output by the adult trematode. Thus, when mice infected with *Schistosoma mansoni* were placed on a low protein diet (12·8 per cent. protein) worms became stunted, developed abnormal reproductive organs and produced fewer eggs than controls fed on a diet containing 20·2 per cent. protein, although the actual growth rate of the experimental mice was unaffected.[74] Comparable experiments have not been carried out on the effects on trematodes of varying the carbohydrate or lipid content of the host diet.

That certain growth substances are likely to be important for normal metabolism in trematodes, as in other organisms, may be predicted on theoretical grounds, but the nature of these substances is unknown. Whether certain amino acids or vitamins are

' essential ' is, in general, unknown.   Accurate information is likely to be forthcoming only when *in vitro* methods are perfected. These in turn depend basically on nutritional data; so the worker here faces a dilemma.   The question of nutrition is further considered when the problems of attempting to grow trematodes *in vitro* are discussed (Chapter 9).

5: The Biology of the Egg

General Account

When the general ecology of trematodes was being considered, it was stressed that adults are found within the hosts only in those habitats which have access (directly or indirectly, e.g. schistosomes) to the outside world; the eggs, in general, then are adapted to external environmental conditions. Since aquatic molluscs act as intermediate hosts for most species, the adaptations are generally those related to an aquatic existence. An exception, however, is *Dicrocoelium dendriticum*, whose early larval stages develop in a terrestrial snail. Trematode eggs are not only adapted for an aquatic environment, but the majority of them are also adapted to hatch under the influence of external environmental stimuli, such as light; they may also possess mechanisms which effectively inhibit premature hatching within their definitive hosts. Occasionally (e.g. *Parorchis acanthus*) eggs may hatch while still *in utero*, although even then the possibility that light is the stimulating influence cannot be ruled out. In the case of *Dicrocoelium*, mentioned above, hatching occurs only when the egg is ingested by the molluscan host.

Most trematode eggs have a well-developed egg-shell or *capsule* enclosing the ovum and a number of vitelline cells; the formation of the egg-shell is discussed more fully below. Most eggs are operculate without spines and filaments. Opercula are lacking in some trematodes, notably the Schistosomatidae, and members of this family may have a short lateral or terminal spine.

The Formation of the Egg

Reproductive anatomy

To understand how the egg and its shell or capsule is formed it is necessary to mention briefly the salient features of the trematode reproductive system (figs. 3 and 10).

Male. The male system presents no unusual features, being based on the platyhelminth plan of paired testes, vasa efferentia, a vas deferens, a seminal vesicle, a ductus ejaculatorius and a cirrus. The ductus is often provided with glands whose physiological activity is unknown.

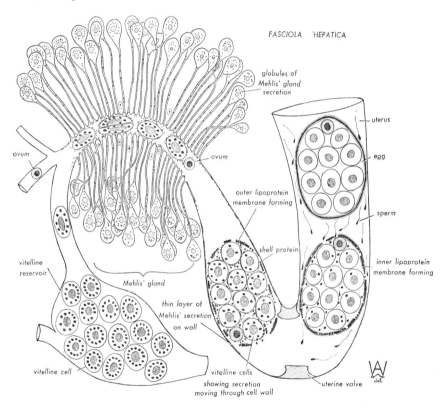

FIG. 10. Mechanism of egg-shell formation in a digenetic trematode; based on *Fasciola hepatica*. (after Clegg, 1965)

Female. The female system consists typically of a single ovary, an oviduct, a receptaculum seminis, paired vitelline glands (=vitellaria), an ootype (a chamber in which the eggs are formed) and a group of glands collectively known as Mehlis' gland. A canal, *Laurer's canal*, leads from the oviduct to open to the dorsal surface and appears to be homologous with the copulation canal of cestodes; it is absent in some species.

Fertilization

The spermatozoon. The morphology of the sperm of the Platyhelminthes is, in general, somewhat aberrant compared with that of other invertebrates.[97] Many turbellarians have more than one flagellum and it is interesting to note that the sperm of *Fasciola* and *Dicrocoelium* are biflagellate and that of *Haematoloechus* may be multiflagellate, in the spermatid stage.[129] The tail (= shaft) of the spermatozoon is formed by these flagellae uniting with a cytoplasmic extension of the spermatid (termed the ' middle snook ' !) during a later stage of development. Electron microscope studies have confirmed that the sperm flagella have the standard pattern of other flagella and sperm tails, being typically composed of two central filaments and nine peripheral doublet filaments.[122]

A mature spermatozoon is about 300–330 μ long and the thread-shaped nucleus at one end is about 50 μ long.[129] The cytoplasmic shaft of the spermatozoon does not taper progressively towards the tip. Following down from the nucleus it gradually becomes very thin and about 90 μ from the end it suddenly becomes thick again. This point may have some particular physiological significance, for, although waves of undulation as a rule travel towards the nucleus at this point, something may happen to change the direction of the waves in the proximal part of the shaft.

Mechanism. Self fertilization probably can occur in trematodes but is unlikely to be common as a number of flukes usually occur together. In general, copulation occurs by the cirrus of one fluke being thrust into the terminal portion of the uterus (often called the ' metraterm ') of another fluke. Instances have also been recorded of a cirrus being inserted into the opening of Laurer's canal and fertilization may take place in this way in some species.

Spermatozoa are stored in the receptaculum seminis and released as required. Nothing appears to be known regarding the physiology of fertilization and whether or not a coordinating hormonal system is involved. Fertilization takes place in or near the ootype and only a single sperm is involved; there is no evidence of polyspermy.[256] Some workers believe a secretion from Mehlis' gland acts as a gamone and activates the spermatozoa before fertilization. If this is so, Mehlis' gland may be found to have a number of functions, for it also takes part in the formation of the lipoprotein envelopes of the egg-shell (see below).

Egg-shell formation: General account

The formation of an egg in trematodes is a complex process in which a series of beautifully controlled reactions result in the release to the outside world of an ovum packed in a container—the egg-shell—made of a resistant tanned protein. This shell is further covered by a lipoprotein layer and provided with reserve material in the form of the residue of the vitelline cells from which the shell-precursor material has already been used. The cytochemistry and cytology of the egg-forming process have been reviewed in some detail.[55, 292]

The mechanism whereby the egg is formed in the majority of trematodes is shown diagramatically in fig. 10 and appears to be as follows.

Ova are released periodically from the ovary and pass down the oviduct; a sphincter muscle, the *oocapt* (=ovicapt), probably controls the release of ova. Simultaneously with this process, a number of mature vitelline cells (approximately 30 in *Fasciola*, 30–40 in *S. mansoni*[118]) are released from the vitellaria; a sphincter muscle similarly controls the release of these cells. An ovum, with its groups of vitelline cells, passes into the ootype and is subjected to the influence of the secretion from Mehlis' gland. At this point the vitelline cells secrete globules of shell precursor which coalesce to form a semi-liquid shell. This encloses the ovum and the remains of the vitelline cells, which serve as yolk reserves in the completed egg. The shape of the egg is apparently determined by the shape of the uterus. This may be particularly well seen in *S. mansoni* (fig. 11) where one pole has a spine.

The bulk of the egg-shell material thus comes from discrete globules in the vitelline cells. That these globules represent the shell precursors is clear from the fact that they give positive results with cytochemical tests for proteins, phenols and phenolase— the materials from which the shell is formed (see below). Many of these tests give brightly coloured end-products, and, when applied to whole worms, enable the vitellaria and the newly formed eggs in the uterus to be selectively stained in a remarkable manner (plate I, fig. B). Thus, a cytochemical approach to trematode morphology is possible, and a number of simple techniques have been developed.[150]

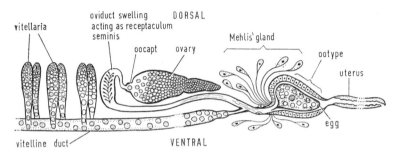

FIG. 11. Morphology of female genitalia of *Schistosoma mansoni*. (after Gönnert, 1955)

The exact rôle of Mehlis' gland in egg-shell formation is uncertain. In some species it appears to contain two types of gland cells producing mucus and lipoprotein secretions respectively. The mucus secretion probably serves to lubricate the uterus for the passage of eggs or it may stimulate sperm activity. In *F. hepatica*[53] the lipoprotein secretion lines the uterus and forms a definite lipoprotein membrane which covers the surface of the vitelline cells and serves as a template on which the shell material is deposited. This internal membrane is believed to be homologous with the vitelline membrane. A further lipoprotein membrane is formed on the *outside* of the shell.[53] Chemical analysis of the lipoprotein forming these membranes indicates a content of 61 per cent. protein and 1 per cent. carbohydrate (including 0·17 per cent. hexosamine). The bulk of the lipid has been found to be cholesterol and triglyceride, only small quantities of phospholipid being

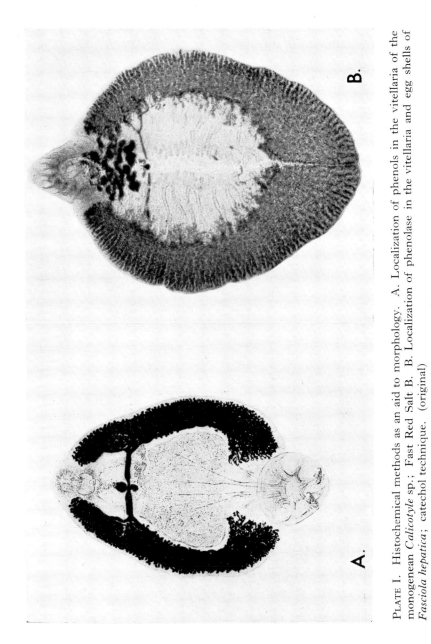

PLATE I. Histochemical methods as an aid to morphology. A. Localization of phenols in the vitellaria of the monogenean *Calicotyle* sp.; Fast Red Salt B. B. Localization of phenolase in the vitellaria and egg shells of *Fasciola hepatica*; catechol technique. (original)

present. The low concentration of phospholipid may explain the fact that these membranes have so far proved difficult to fix for study by electron microscopy.[55]

The newly formed egg with semi-soft shell is passed along the uterus, becoming tanned and hardened as it does so; gradually it loses its ability to react to histochemical tests for phenols, phenolase and basic proteins.

The egg-shell is thus envisaged as being a thick tanned protein wall covered on each side by a thin lipoprotein membrane. The lipoprotein membranes are permeable to water and probably impart semi-permeability to the egg capsule, since salts inside the egg are retained.[53] In the operculate egg of *Fasciola* the operculum is formed by a remarkable process at the end of the egg containing the ovum.[119] The ovum apparently pushes out pseudopodia towards the egg-shell during its process of formation exactly at the point of rupture of the operculum. The result is that the egg-shell is weakened at this point and so readily fractured on hatching.

A plexus consisting of four or five nerve cells has been found in the region of the yolk duct and may coordinate the activities of the oviduct, vitelline ducts and Mehlis' gland. Another plexus, consisting of only two nerve cells, lies near the upper uterus and may serve in the control of a uterine valve (which prevents the backward flow of eggs into the uterus) and the lower ootype in general.[119]

An interesting and unique modification of the above mechanism is found in *Syncoelium spathulatum*.[56] In this species the phenolic and protein components of the egg-shell materials are apparently released from glands in the wall of the uterus and are deposited on a thin membrane produced from Mehlis' gland. In *S. mansoni* too, the uterus may produce a secretion necessary for egg-shell formation.[118]

Chemistry of egg-shell formation

General theory. The egg-shell in the majority of trematodes is composed of a stable, highly resistant type of protein termed sclerotin.[55, 292, 301] This is a ' tanned ' protein and one which occurs in structural proteins in widely divergent groups in the invertebrates and vertebrates. Thus, the insect cuticle and the elasmobranch egg case, for example, consist largely of a sclerotin-type of protein. The sclerotinization or 'tanning' of this kind of protein is accomplished by means of an *o*-quinone derived enzymatically from an *o*-phenol in the presence of oxygen. The process is sometimes referred to as 'quinone tanning'. The quinone reacts with free NH_2 groups on adjacent protein chains

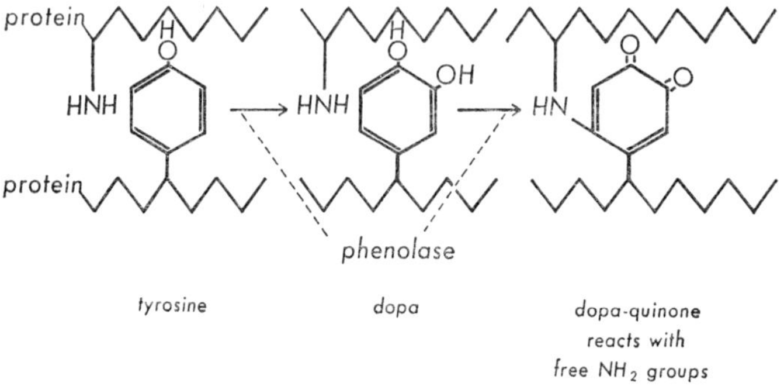

FIG. 12. Chemical reactions involved in quinone tanning.

to form a cross-linked stable protein (fig. 12). These reactions may be summarized as follows:

$$\text{(I)}\quad \begin{array}{l} \text{phenol} \xrightarrow{\text{phenolase}} \text{quinone} \searrow \\ \qquad\qquad\qquad\qquad\qquad\qquad \text{`tanned' protein} = \text{sclerotin.} \\ \text{protein} \longrightarrow \text{protein} \nearrow \end{array}$$

This is probably the system followed in the majority of trematode egg-shells, but in some species the oxidizing enzyme involved, phenolase, appears to be lacking and some other system, as yet undetermined, may operate. An alternative explanation is that the enzyme may be destroyed by the methods used to identify it! In *S. mansoni*, phenolase can be demonstrated after two hours' fixation in 70 per cent. ethanol but not after twenty-four hours' fixation.[55] Typically, trematode eggs are almost colourless when formed first in the early part of the uterus, but become dark yellow or reddish-brown as the tanning proceeds; some eggs, however, are almost colourless even when hardened.

Phenolic substances in egg-shell. Attempts to identify the phenolic materials concerned in egg-shell formation have not so far been successful.   The identification is likely to prove an extremely difficult task, for even in insects—where much more abundant material is available—the identification has taken many years of patient work.   In the blowfly larva *Calliphora* for example, it has been shown that labelled tyrosine injected into the larva just before pupation is converted into n-acetyldopamine,

(II)

$$\text{HO}\!-\!\!\!\overset{\displaystyle CH_2\!-\!CH_2}{\underset{\displaystyle HN\!-\!COCH_3}{\bigcirc}}$$

which is then oxidized enzymatically to a quinone.

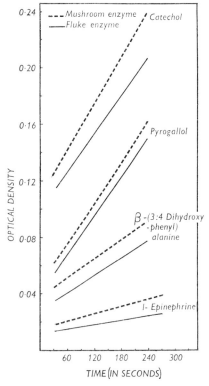

FIG. 13.   Comparison of the activity* of phenolases from *Clonorchis sinensis* and mushroom with different substrates, in phosphate buffer at *p*H 6·8.   (after Ma, 1963)

* Activity measured in terms of optical density of products of oxidation.

In *F. hepatica* a minute quantity of an indolic compound has been extracted from the vitelline glands[104] and may prove to be the serotonin-like substance possibly involved in nervous transmission.

Failure to detect free phenolic tanning materials suggests that tanning may occur by the oxidation of the tyrosine bound to protein as shown above, a process which has been referred to as

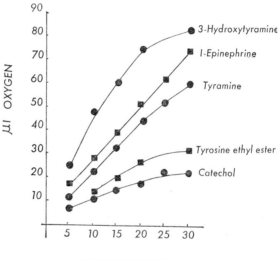

FIG. 14. Activity of phenolase from *Fasciola hepatica*, estimated in terms of oxygen consumption, with various substrates. (after Mansour, 1958)

' *autotanning* '.[55] Tyrosine has been detected in the vitelline cells of *F. hepatica*[16] but, without rigorous chemical examination, this is not enough evidence to prove that autotanning occurs in this species.

Properties of phenolase. Phenolase (o-diphenol: O_2 oxidoreductase, polyphenol oxidase, catechol oxidase) are complex copper-containing enzymes which may act on monophenols as well as diphenols.[55] Phenolases, which are presumably involved in egg-shell formation, have been isolated from *Clonorchis sinensis* and *F. hepatica* and their properties examined.[200, 209] The enzyme from *C. sinensis* resembles, in its general properties, the comparable enzyme isolated from *F. hepatica* or from mushrooms, but it is slightly less active than the latter and shows other differences. Catechol and pyrogallol are good substrates but β-(3:4-dihydroxyphenyl) alanine and 1-epinephrine somewhat poorer (figs. 13, 14). The drug serotonin has been found to reveal differences between phenolases from different sources.   Thus, it inhibits the action of

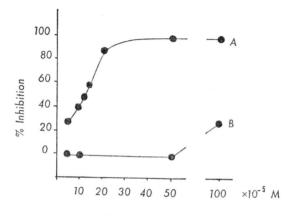

Sodium diethyldithiocarbamate

FIG. 15.   Inhibition of phenolase from *Fasciola hepatica* by sodium diethyldithiocarbamate; substrate-tyramine. A. No CuSO$_4$ added, 100 per cent. inhibition.   B. Activity restored by the addition of 10^{-5}M CuSO$_4$. (after Mansour, 1958)

phenolase from *F. hepatica* on substrates of phenolamines and catecholamines; the action of phenolase from mushrooms is, however, unaffected.[209]   Serotonin also inhibits the action of phenolase from *C. sinensis* on 1-epinephrine but not on phenols,[200] thus indicating that the phenolase from *C. sinensis* and *F. hepatica* has the same specificity for catecholamines.   Sodium diethyldithiocarbamate, a drug which probably forms colour-complexes with the copper, strongly inhibits phenolase activity in both trematodes and mushrooms; enzyme activity may be restored, however, by adding copper ions in the form of 10^{-5}M CuSO$_4$(fig. 15).

Embryonation and Hatching

Factors influencing embryonation

Eggs pass to the outside via the three main body wastes—sputum, urine or faeces. In some cases, e.g. the Schistosomatidae, or those families whose eggs hatch on ingestion by a snail, embryonation occurs within the body and eggs are ready for hatching when laid. In many species, however, eggs are unembryonated when laid, and remain so until they encounter suitable environmental conditions in the outside world. Many physico-chemical factors, known to influence biological reactions generally, influence the embryonation of the egg, but humidity, temperature and oxygen tension are particularly important. A knowledge of these is clearly important for an understanding of the ecology of trematodes.

Influence of faeces. The presence of faecal material appears to have a markedly inhibitory effect on embryonation. Thus, although in moist or wet faeces eggs of *Fasciola* can remain viable for long periods (fig. 16) and may undergo partial development, they do not hatch in faeces—an effect which might be due to a number of factors, e.g. competition for oxygen by microflora, presence of toxic substances or enclosure by films of bacteria.[268]

Influence of temperature. The effect of temperature on the embryonation time of *F. hepatica* eggs is shown in fig. 16; comparable results are obtained with *F. gigantica*.[162] The critical temperature, below which embryonation does not occur, is $9 \cdot 5 \pm 0.5°$ C.[268] in *F. hepatica* but is 6° C. in *Fascioloides magna*.[32] Within the range 10–30° C. the rate of embryonation increases with temperature. Thus, at 10° C. embryonation of *F. hepatica* eggs takes 23 weeks but at 30° C. is completed in 8 days. Above 30° C., development is increasingly inhibited and at 37° C. does not occur at all. Eggs incubated at 37° C. and then transferred to a lower temperature will incubate but the mortality increases, the longer eggs remain at 37° C.[268] At 37° C. eggs are killed within about 24 days. Temperatures below freezing rapidly kill trematode eggs. Eggs of *Fasciola gigantica*, for example, at $-3°$ C. and $-5°$ C. are killed within 23 and 17 days respectively,[162] and eggs of *F. hepatica* are rapidly killed near the same temperatures, but may survive for some time under 30–40 cm. of snow.[333]

Influence of moisture. It seems that few precisely controlled experiments on the effect of moisture on trematode eggs have been carried out. It is well known, however, that eggs desiccate

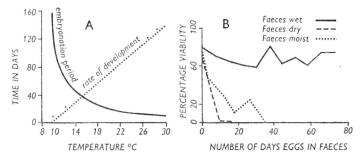

Fig. 16. A. Influence of temperature on the development of the egg of *Fasciola hepatica*; B. Viability of eggs of *Fasciola hepatica* under different moisture conditions at summer temperatures. (after Rowcliffe and Ollerenshaw, 1960)

readily and that water is required for embryonation. Eggs on soil, however, will develop without the presence of free surface water, provided that the soil is saturated with water.[268] In wet faeces eggs remain viable for more than fifty days but do not

TABLE 13

Influence of oxygen tension on viability and hatching time of Fasciola hepatica *eggs*

(After Rowcliffe and Ollerenshaw[268])

	Aerobic*	Open to air	Anaerobic
O₂ in gm. per litre	0·03288	0·03207	0·00405
No. of days for 50% hatch	29	43	115
Percentage dead at time of 50% hatch	38	53	39

* With air bubbling through the culture.

survive for more than several weeks in dry faeces, or more than a month in ' moist ' faeces.

Influence of oxygen tension. Eggs of *F. hepatica* show little variation in mortality at different oxygen tensions but those in aerobic conditions hatch in one-fifth of the time taken by those at a lower oxygen tension (table 13).

Influence of pH. Eggs of *F. hepatica* will embryonate and hatch within a *p*H range of 4·2–9·0 but above about *p*H 8·0 development is prolonged.[268]

Hatching

General comments. It is clearly of major survival value for a parasite that the hatching processes of the egg should be stimulated

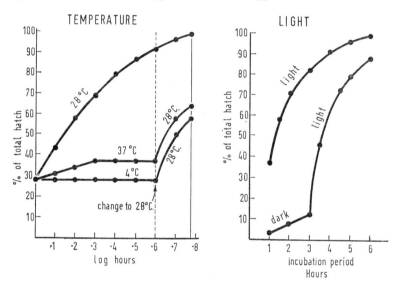

Fig. 17.   Influence of temperature and light on hatching of eggs of *Schistosoma mansoni*. (after Standen, 1951)

by environmental conditions which also provide it with a reasonable opportunity for infecting the intermediate molluscan hosts. At the same time the process should be *inhibited* by conditions within the definitive host, so that hatching shall not occur prematurely within the definitive host.

Factors affecting hatching. In species such as *S. mansoni*, in which the egg is fully embryonated when laid, temperature, light and salinity have a marked effect on hatching. Thus, at 28° C. (fig. 17) eggs hatch readily; but low (4° C.) and high temperatures (37° C.) inhibit the process, although hatching is restored after eggs are returned to 28° C. from such temperatures. In this species hatching is almost completely inhibited by o·6 per cent. NaCl and extensive hatching does not occur until a dilution to about o·1 per cent. NaCl is reached; this mechanism ensures that eggs in blood, gut contents or urine will hatch only on dilution with water. Stimulation of hatching in such conditions appears to be related to an osmotic effect on the activity of the miracidium. Thus, in the egg of *S. mansoni* the flame cells are inactive in saline but commence flickering a few seconds after being placed in distilled

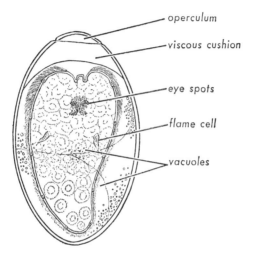

operculum

viscous cushion

eye spots

flame cell

vacuoles

FIG. 18.   Embryonated egg of *Fasciola hepatica.* (original)

water. The relative importance of osmotic pressure, temperature or light seems to differ greatly between different species. In some species (e.g. *Fascioloides magna*) substantial hatching (80 per cent.) may be brought about by subjecting the eggs to an atmosphere of N_2 or by reducing the atmospheric pressure to 5–15 cm. Hg— both conditions unlikely to be encountered in nature.[102] In most species, however, light has a marked stimulatory effect on hatching, the mechanism evoked probably being similar to that described for *Fasciola* (see below).

Mechanism of hatching. The mechanism of hatching has been worked out in some detail for *F. hepatica*.[266, 267] An embryonated egg of *F. hepatica* is shown in fig. 18. The miracidium is enclosed in a *vitelline membrane* which is normally not readily seen.   At the opercular end lies a *viscous cushion*, a mass of colloidal substance just beneath the operculum but covered by the vitelline membrane. The anterior end of the miracidium is directed towards the viscous cushion and the miracidium is partly surrounded by two fluid-filled vacuoles which occupy almost all the remainder of the egg cavity.

Under the influence of light, eggs hatch within a matter of minutes.   There is some experimental evidence to suggest that light stimulates the miracidium to release a proteolytic ' hatching enzyme' which attacks a proteinaceous 'cement' material, believed to hold the operculum in place, and frees the operculum.[266]   The

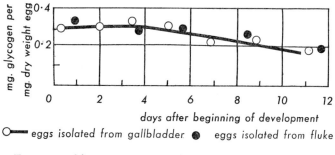

FIG. 19.   Glycogen content of eggs of *Fasciola hepatica* during embryonation at 25° C.   (after Hortsmann, 1962)

operculum is released instantaneously from all but one point of contact with the shell.   Just before hatching, the viscous cushion expands to about twice its original size probably as the result of a change from a contracted gel to an expanded sol.   This swelling may be related to osmotic effects due to changes in the permeability of the vitelline membrane brought about by the action of the hatching enzyme.   When the operculum is released, the vitelline membrane ruptures and the semi-fluid cushion material flows out and disperses in the water.   At the same time, the activity of the miracidium increases, its cilia beat rapidly and the miracidium escapes through the opening.   The rôle of the viscous cushion in hatching is unexplained;   it appears to take no active part in releasing the operculum or activating the miracidium.

If the existence of a ' hatching enzyme ' is confirmed, it presents a most unusual and interesting biochemical phenomenon which has been little investigated. There is some evidence that the violet and blue wavelengths of the spectrum appear to be the essential part of the light stimulus.[259]

The hatching mechanism on non-operculate eggs, such as those of *Schistosoma*, has not been worked out; neither has it been examined in the case of the numerous species, such as *Dicrocoelium dendriticum*, in which the egg hatches only on ingestion by a

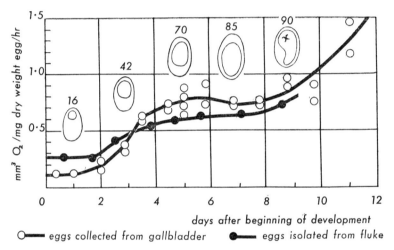

FIG. 20. Oxygen consumption of egg of *Fasciola hepatica* during embryonation at 25° C. (after Horstmann, 1962)

snail. Presumably, in this case the host provides some stimulus (e.g. level of CO_2) which stimulates the release of a hatching enzyme within the egg; or it may be simply that the digestive enzymes of the snail attack the opercular seal directly.

Metabolism

The general metabolism of the egg is poorly known and has been examined for only a few species. Like other invertebrate eggs, those of trematodes are rich in enzymes; and acid and alkaline phosphatases, 5-nucleotidase, adenosine triphosphatase, aminopeptidase, an acetylcholine esterase, and a non-specific esterase have been detected in the eggs of *S. mansoni*.[4, 246] Carbohydrate reserves are present in the form of glycogen, and in *F. hepatica* the glycogen content falls from 32 per cent. (dry wt.) to about 15 per cent. (fig. 19) at the time of hatching.[133]   During the early phase of development the oxygen consumption is less than 0·3 mm.³/mg. (d.w.) hr. but increases as development proceeds (fig. 19).   This is a pattern similar to that found in other invertebrate embryos and clearly reflects the higher demands of the synthetic phases through which the egg is passing.   The fact that oxygen is required for embryonation does not necessarily mean that it has a predominantly aerobic metabolism.   This may prove to be the case, or it may simply mean that oxygen is required to synthesize some essential growth metabolite.   It will be recalled that adult trematodes have a predominantly anaerobic metabolism, which may be carried back to the larval stages.   The eggs of some species are embryonated when laid and hence may require only low oxygen tensions for hatching.

The metabolic pathways during embryonation have not been worked out and apparently nothing is known concerning the fat or protein metabolisms.

6: Intramolluscan Stages

Miracidia

Anatomy

It is not intended to give here a detailed account of the anatomy of a miracidium, the main points of which are shown in figs. 21-24; but attention may be directed to certain features. A freshly emerged miracidium is generally pyriform or bullet-shaped. The body is covered by flat ciliated epidermal plates, beneath which is a well-developed sub-epidermal musculature (fig. 22). A miracidium swims actively at a speed of about 2 mm. per sec. at ordinary temperatures. The anterior end bears a blunt retractable *apical papilla* or *terebratorium* which is devoid of cilia but bears two pairs of filaments and five pairs of gland openings. At their bases the filaments have nerve cells which make contact with the cerebral ganglion; it has been suggested that the filaments are concerned with chemoreception.[83]

Of particular interest is the *apical gland*, a prominent saccular structure in the anterior third of the body, which contains four nuclei. This is often termed a ' primitive gut ' but is believed to be a gland which probably secretes histolytic enzymes (see p. 83). On each side of the apical gland are the so-called ' penetration glands ' opening by narrow ducts at the base of the apical papilla (fig. 24). Additional gland cells have been described in some species. In general, miracidia are well supplied with what appear to be sense organs. Thus, most miracidia have a pair of eyespots, sometimes fused together. Each eyespot is a cup, composed of pigment granules and containing one or two lenses. The nature of the pigment has not been determined but, as in the cercaria (p. 95), is likely to be melanin.   Immediately behind the eyespots is a mass of cells with some fibres, representing the ' brain '. Between the first and second series of epidermal cells are located a pair of knob-like papillae (=*lateral processes* or *papillae*) from which nerve fibres to the brain have been traced in some species.   These are believed to be sensory (possibly chemoreceptive[83]) but may be glandular.   *Radial papillae* are found anterior to the lateral processes, one at the base of each epidermal plate of the first ring of plates.   Other series of papillae in some species have been described.

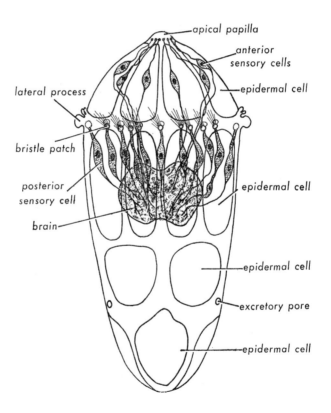

FIG. 21.   Miracidium of *Schistosomatium douthitti* to show superficial anatomy. (after Price, 1931)

The interpretation that these cells are sensory in function is based on the fact that each cell terminates in the body surface and the brain. The number of sensory cells appears to vary widely, even in the same species. Bristle patches are associated with some sensory cell terminations (fig. 23).

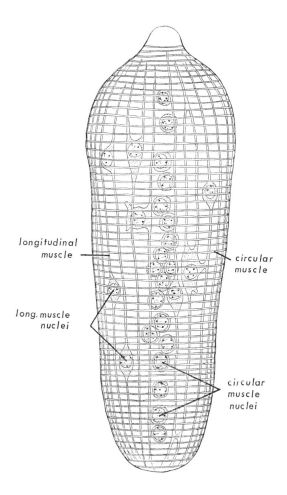

longitudinal
muscle

circular
muscle

long. muscle
nuclei

circular
muscle
nuclei

FIG. 22.   Muscular system of miracidium of *Posthodiplostomum cuticola*. (after Dönges, 1964)

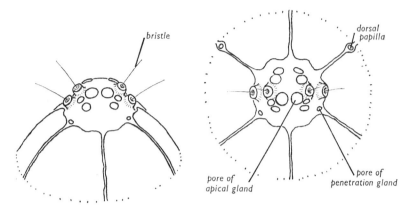

FIG. 23.   Terebratorium of miracidium of *Posthodiplostomum cuticola*.   (after Dönges, 1964)

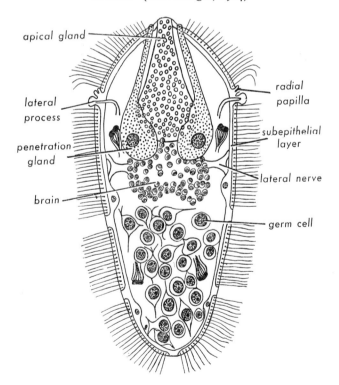

FIG. 24.   Miracidium of *Schistosomatium douthitti* to show internal anatomy.   (after Price, 1931)

General responses

Miracidia, in general, probably exhibit behaviour patterns somewhat similar to those of their molluscan hosts. Many show phototactic, thermotactic and geotactic responses, the intensity of which varies from one species to another. In the case of *S. mansoni*, negative geotaxis can be shown experimentally to be more dominant than phototaxis (fig. 25). This is, however, an over-simplification of the problem, since each response is to some extent dependent on other factors. By means of the apparatus

TABLE 14

Effect of temperature and light intensity on phototaxis in miracidia of Schistosoma japonicum
(Data from Takahashi et al.[315])

LUX/°C.	15	18	20	22	23	24	25	26	28	30	34
4500	+	−	−	−	−	−	−	−	−	−	−
2500	+	±	−	−	−	−	−	−	−	−	−
2000	+	±	±	−	−	−	−	−	−	−	−
1000	+	+	±	±	±	±	−	−	−	−	−
500	+	+	+	+	+	+	±	±	−	−	−
250	+	+	+	+	+	+	+	+	±	−	−
100	+	+	+	+	+	+	+	+	+	−	−
50	+	+	+	+	+	+	+	+	+	±	−
25	+	+	+	+	+	+	+	+	+	+	−
10	+	+	+	+	+	+	+	+	+	+	±

shown in fig. 26 the miracidia of *S. japonicum* has been found to exhibit positive phototaxis to any light intensity at 15° C., whereas at 20° C. they show positive reactions to 2000 Lux or less, at 25° C. to 500 Lux or less, and at 30° C. to 50 Lux or less (table 14). Under a given intensity of light, the velocity of miracidia induced by the phototactic response is linearly relative to the water temperature. With a light intensity of 2500 Lux the relationship between the two can be expressed by

$$y = 0 \cdot 075x - 0 \cdot 315$$

where y is the velocity (mm./sec.) and x the temperature (° C.).[315] Miracidia show a strong negative geotactic response which is disturbed at temperatures above 20° C. by an intensity of more than 5000 Lux.[315] At 15° C. the negative geotaxis of this species

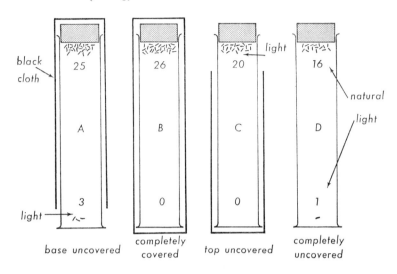

FIG. 25. Geotactic and phototactic responses of *S. mansoni* miracidia. The numbers indicate the number of miracidia found in samples withdrawn by hypodermic syringes from the top and bottom of cylinders illuminated in different ways. (based on data from Chernin and Dunavon, 1962)

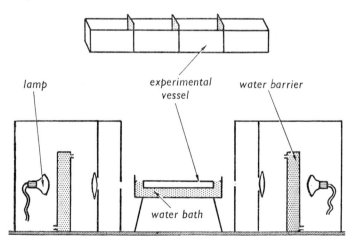

FIG. 26. Apparatus used to examine the phototactic and geotactic responses of the miracidia of *Schistosoma japonicum*. (after Takahashi, Mori and Shigeta, 1961)

is less dominant than the negative response to light, and so larvae move to the bottom. The behaviour of *Oncomelania nosophora*, the molluscan host of *S. japonicum*, very closely follows that described above for the miracidium, and this parallelism must play a major part in establishing the infection of the snail.

The geotactic and phototactic responses of miracidia may be used by means of the side arm flask, to separate them from a suspension of eggs in water. The macerated tissue containing eggs is placed in the flask, the bulb of which is covered with dark material and the side arm illuminated. The miracidia swarm into the side arm and may be removed.

Survival of miracidia is, of course, markedly influenced by temperature. The ' half-life ' of miracidia of *Schistosomatium douthitti* for example, is only 1½ hr. at 35° C., but 11 hr. at 8° C.

Host location

The mechanism whereby a miracidium is brought in contact with its molluscan host has been a matter of controversy for some years. A number of workers have claimed to have observed chemotaxis[90, 91, 95, 163, 361] whereas others could demonstrate no specific attraction.[1,33] The chief difficulty has been to design a suitable experiment to give an unequivocal result.

One series of experiments with the miracidia of *S. mansoni* appears, at first sight, to demonstrate that a chemical stimulus from the mollusc is involved.[91] The apparatus (chemotrometer) used in these experiments is shown in fig. 27. It consists essentially of a central chamber with four side arms connected in turn to four small terminal chambers. The entire inner surface of the apparatus was coated with an inert waterproof lacquer to prevent contamination. One snail was placed in each of two of the terminal chambers and sham snails of inert aquarium cement were placed in the other two terminal chambers to equalize the amount of light reflected into the central chamber through each arm; selection of the appropriate chambers was randomized in each experiment. The entire apparatus was filled with ' conditioned ' water. It was then allowed to stand for one hour or more to allow any substances produced by the snails to diffuse with the water. The shell of each snail was routinely cracked to prevent it from moving into the central chamber in which 100–200 miracidia were placed. After one hour, 60–80 per cent. of the miracidia had entered the side arms; each arm was then stopped and drained, and the number of miracidia counted. Results are shown in table 15. In nine valid experiments, 569 out of 975 miracidia entered the arms with real snails and only 406 entered the arms containing the sham snails. The differences were statistically highly significant. A shorter series of four experiments gave a

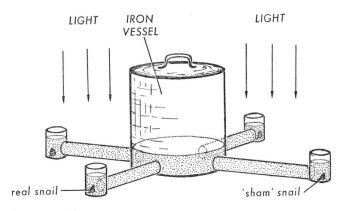

FIG. 27. Chemotrometer used by Etges and Decker (1963) to test for chemotaxis of miracidia of *S. mansoni* for various snail species. For details see text. (original)

somewhat similar result. With two other non-host snail species, however, no significant attraction was found. The experiments with cracked snails are open to criticism on the grounds that the leaked blood could act as the attractant, for amino acids and fatty acids appear to serve as chemical attractants.[154, 204, 205]

The nature of a possible chemotactic substance in a snail which may attract the miracidia of *S. mansoni* has not been determined; it may be present in the mucous secretion. Alternatively a physical condition, such as pH or pCO_2—related to the presence of snails or substances secreted by them—may be involved. Experiments have shown that miracidia of *Paragonimus ohirai* show positive responses to three species of snails or to a homogenate of one species or to an amino acid mixture encapsulated in cellophane.[154] The miracidia in this case were attracted to each species of snail used, independent of host suitability. Further evidence[60] that the miracidia are influenced by snail tissue is the fact that in pure water miracidia tend to swim in straight lines, as shown by

TABLE 15

Test for chemosensitivity of miracidia of Schistosoma mansoni *as indicated by experimental distribution in the presence of three species of snails*
(Data from Etges and Decker[91])

Snail Species	No. of Expts.	No. of miracidia started	No. of miracidia with real snails	No. of miracidia with sham snails	X^2	P^2
Australorbis glabratus (shell cracked)	9	1359	569	406	$+27\cdot24$	0·0005
A. glabratus (shell intact)	4	781	342	246	$+15\cdot68$	0·0005
Bulinus sp.	6	1224	401	459	$-3\cdot92$	0·05
Helisoma anceps	1	225	82	114	$-5\cdot22$	0·025

the flying spot microscope, whereas in water containing snail extract they show agitated movements (fig. 28).

These results enable us to speculate on the possible mechanism of host location employed by miracidia. The first phase could be the reaction of the larvae to physical stimuli such as light, gravity,

FIG. 28.   Path of miracidia of *Schistosoma mansoni* as revealed by the flying spot microscope. (after Davenport, Wright and Causley, 1962)

temperature, salinity and pH, some of which have been described above. These tropisms, which are closely allied to those of molluscs, would bring the miracidia into the host environments so that infection now becomes possible. Here the movements become

T.—F

random until the larvae come within chemotactic range of their hosts. The distance over which any chemotactic attraction is effective will depend substantially on the physical conditions in the water governing the diffusion of the substances concerned. Miracidia have been found successful in locating and infecting hosts situated at distances of 86 cm. on the horizontal and at distances of 44 cm. downwards. There is no reason to believe that these represent the limits of miracidial scanning capacity.[46] Other unknown factors, too, may inhibit the attraction, for larvae may sometimes be seen swimming in the immediate vicinity of a snail— even butting against it—without ever attacking it, or conversely attempting to penetrate organisms other than molluscs, such as planaria and oligochaetes.[217] The evidence that miracidia are selectively attracted to snails with a high degree of host 'suitability' is unconvincing. The miracidia of *S. mansoni*, for example, have been shown to penetrate 17 out of 19 snail species exposed to them, but penetrating miracidia were destroyed by a tissue reaction in the non-host species.[314]

Metabolism

The carbohydrate metabolism in *Fasciola* miracidia has been examined by investigating the incorporation of ^{14}C labelled glucose and succinate into metabolic intermediates.[21] Like the adult fluke (p. 42), labelled glucose was found to be incorporated into the glycolytic intermediates, hexose phosphates, phosphoenol-pyruvic acid, alanine and lactic acids. The Embden-Meyerhof pathway of glycolysis thus clearly operated.

In contrast with the adult, however, a Krebs cycle does not appear to occur in the miracidia, although there is some evidence for the existence of two enzymes of that cycle, succinic dehydrogenase and fumerase. Miracidia can use glucose from exogenous sources; the addition of 0·1 per cent. glucose to the water in which miracidia are retained for experimental infections might therefore be considered as a possible means of increasing the rate of infection.

The respiration of miracidia or other aspects of their metabolism appear not to have been studied.

Penetration

It is commonly held that a miracidium enters a snail by boring into the host tissue with a mechanical, auger-like action of its anterior papilla. Whereas there is no doubt that the movements of the miracidia do assist in penetration, the process of entry is probably largely affected by enzyme action.

During the first phase of penetration[62-65] the miracidium becomes attached to the snail either by an adhesive secretion or by means of the suctorial action of the apical papilla. The latter may appear to rotate in some cases and show intermittent, rapid lashing or contracting movements. At the point of attachment, the mollusc tissue is cytolysed and after about fifteen minutes about one-third of the body is embedded. At this stage some miracidia, for no apparent reason, are seen to withdraw the apical papilla and attempt penetration at a new site or swim away. As penetration proceeds, the miracidia of some species shed the ciliated epithelium but others retain it until after penetration is completed. The last final thrust of the larva into the tissue is rapid and the miracidium suddenly disappears into the body of the snail. The whole process takes about thirty minutes.

The rôles played by the various glands are uncertain, and the enzymes concerned have not been identified. The apical gland is thought[360] to secrete the bulk of the tissue-dissolving enzymes involved in penetration, on the grounds that it is empty after penetration. Some workers believe that the so-called ' penetration glands ' may be secreting material which forms the cuticle of the sporocyst. In some species the eggs are ingested directly by the snail and hatching occurs in the gut; little appears to be known of the mechanism of hatching or penetration in such cases.

Sporocyst and Redia: General Physiology

Development

In most cases, a miracidium becomes transformed into a mother sporocyst at the site of penetration, little or no migration taking place. This site is generally the mantle tissue, the foot or the tentacles. The ciliated cells are lost, if they have not already been shed during penetration, and the larva becomes an elongated sac which contains germinal cells. The latter give rise either to daughter sporocysts or rediae, depending on the developmental pattern. When fully developed, rediae or daughter sporocysts escape from the mother sporocyst—in most cases through a birth pore—and make their way to the hepato-pancreas, or its immediate vicinity, where they give rise to cercariae. In a few species (e.g. *Parorchis acanthus, Philophthalmus gralli*) precocious development

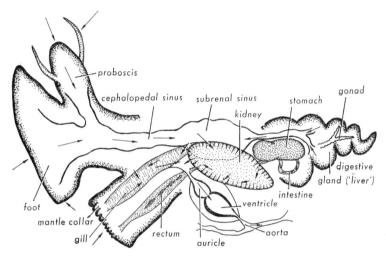

FIG. 29. Anatomy of a melanid snail (*Tarebia granifera mauiensis*) and the possible route of the newly liberated young mother redia to the heart of the snail following miracidial penetration. The heart is reached in about 45 minutes. The miracidium is unusual in containing a young mother redia within its body cavity. (after Alicata, 1962)

occurs and mother rediae develop within miracidia. In such cases a rapid migration may take place; rediae of *P. gralli* reach the heart within forty-five minutes[2] (fig. 29).

Comparatively little is known about the physiology of the intra-molluscan stages. The rate of development of sporocysts and rediae depends on the nutritional state of the mollusc host and the physical conditions (especially temperature) under which it is maintained. Thus, well-fed specimens of *Lymnaea truncatula* were found to measure in the range 0·82–1·22 cm. and contain 133–344 rediae and 952–2275 cercariae of *F. hepatica*. Starved snails, on the other hand, measured 0·62–0·74 cm. and only 30–215 rediae and 9–480 cercariae were found.[159] Even in well-fed snails there will be competition for available nutriment; for most species there is a limit to the number of larvae that can be supported, beyond which cercarial development will be retarded. In *F. hepatica*, about 40 rediae appears to be the most ' economic ' infection; beyond this number, development is considerably retarded. The effect of temperature follows that usual for most biological processes. In most species, development appears to be completely retarded below about 5° C.

Nutrition and general metabolism

Feeding mechanisms. Sporocysts lack a mouth and must therefore absorb all their nutriment through the body wall. In most cases, the presence of the sporocyst appears to stimulate histolysis of the hepato pancreatic cells and therefore frees materials for absorption. Rediae, on the other hand, possess a simple gut and appear to be able both to ingest cells and to absorb material through the body wall.

The mechanical damage caused to hepatopancreatic cells by rediae of certain species (e.g. *Echinoparyphium* sp. in the snail *Helisoma trivolvis*) is much more severe than that resulting from lysis of cells caused by sporocysts. The process of cell destruction and removal appears to be brought about through direct ingestion by rediae. This interpretation is borne out by the presence of large quantities of cellular debris found in the intestinal caecum of the redia and is further supported by the fact that where the heaviest concentrations of rediae occur, there the most severe cellular damage is found. Rediae thus acquire their carbohydrate and other nutrient material by actively digesting hepatopancreatic cells,[36] although this does not rule out the possibility of direct absorption through the body wall, as in the case of sporocysts. That the redial intestinal caecum is functional is further borne out by the fact that high concentrations of acid and alkaline phosphatase (see p. 38) occur there.

Respiration. The amount of oxygen available to the larvae in the majority of mollusc hosts is probably quite small. An exception may be species such as *Gorgodera amplicava*, which occurs in the gills of lamellibranchs, a site where presumably the level of oxygen tension is high. As adult trematodes do, rediae and sporocysts also use oxygen if it is offered to them. The rediae of *Himasthla quissetensis* efficiently regulate the respiration rate well down to 0·5 per cent. O_2 and even at this low level the Q_{O_2} was 61 per cent. that at the 5 per cent. level. This is in marked contrast with the respiration of the cercaria where the Q_{O_2} in 0·5 per cent. O_2 was only 17 per cent. of the rate in 5 per cent. O_2.[336]

The effect of temperature on respiration of larval trematodes is somewhat unexpected. In biological systems the effect of temperature is examined quantitatively by reference to the Q_{10} (the ratio of the rate of the reaction at two temperatures differing by 10° C. i.e. $\left(\dfrac{\text{Rate at } (t+10)°\text{ C.}}{\text{Rate at } t°\text{ C.}} \right)$. The Q_{10} of the respiration of the sporocyst of *H. quissetensis* (fig. 30, and table 16) is only 1·8 even

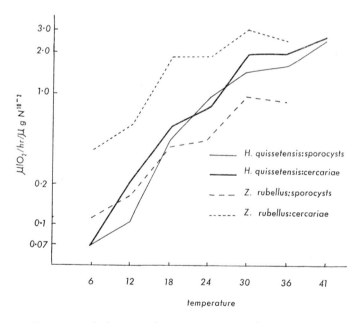

Fig. 30.   Influence of temperature on the oxygen consumption of sporocysts and cercariae of two species of trematodes. (based on Vernberg, 1961)

at 30–41° C. Since, in free-living animals, the Q_{10} tends to be low within the 'normal' temperature habitat range, the Q_{10} of the sporocyst reflects the temperature which would be encountered in the bird definitive host rather than in the molluscan intermediate host. This is an interesting result and may indicate physiological pre-adaptation for the definitive bird host.

Carbohydrate metabolism. Like adult trematodes, sporocysts and rediae also have a pronounced carbohydrate metabolism. The tissue around daughter sporocysts contains less glycogen than the corresponding tissue in *uninfected* snails[35, 42] whereas the contained cercariae are rich in glycogen. The same pattern occurs

TABLE 16

Q_{10} *values for oxygen consumption of larval trematodes*
(Data from Vernberg[335])

	Zoogonus rubellus		Himasthla quissetensis	
Temp.	sporocyst	cercaria	redia	cercaria
6–12	1·9	3·4	2·1	5·5
12–18	4·8	6·7	9·3	4·8
18–24	—	—	3·6	2·9
24–30	3·9	3·9	2·3	4·9
30–41	—	—	1·8	1·5

with rediae. These results may indicate competition between host cells and parasite, rather than the release of amylase by the latter. The presence of parasites at the intramolluscan stage may, however, stimulate excessive breakdown of glycogen in hepato-pancreatic cells. Since glucose has been detected in the sporocyst wall in some species, it is apparent that it can be utilized rapidly by the parasite; for only traces can be detected in the host cells.[44] In some other species glucose does not appear to be utilized. Thus glucose, fucose, mannose and ribose do not increase the respiration rate of rediae of *H. quissetensis* (table 17) and these substances clearly do not act as substrates in this species. On the other hand, the amino acids, proline and glutamic acid, which are known to be glucogenic (i.e. capable of being used for carbohydrate synthesis) increase respiration by 46 per cent. and 20 per cent. respectively.[341] This suggests that, in rediae of this species, carbohydrate synthesis is via these and possibly other amino acids.

Whether a Krebs cycle or a modification of it is adopted by sporocysts or rediae is not clear. In *H. quissetensis*, of three Krebs cycle intermediates tested, pyruvate, ketoglutarate and succinate, only the last named stimulated respiration. Malonate, which competitively inhibits succinic dehydrogenase, has been found to decrease the Q_{O_2} sharply. These results may indicate the operation of a partial Krebs cycle, a situation which occurs in other parasitic groups.

TABLE 17

Utilization of substrates of rediae, cercariae and adults of Himasthla
quissetensis
(Data from Vernberg and Hunter[341])

Per cent. change in O_2 consumption

Substrate	rediae	cercariae	adult
Pyruvate	o	—	—
Ketoglutarate	o	—	—
Succinate	+40	o	+27
Malonate	−24	o	−64
Serine	−12	—	—
Proline	+46	o	+16
Glutamic acid	+20	—	+22
Mannose	o	—	+50
Glucose	o	o	+49
Fucose	o	—	—
Ribose	−11	—	o

Protein metabolism. Only limited work has been carried out on the protein metabolism of trematode larvae. Protein synthesis undoubtedly proceeds at a high rate, for larvae must build not only their own tissues but also those of the individuals of subsequent generations.

As was pointed out earlier (p. 85), the nutritional problems of sporocysts and rediae are somewhat different. Sporocysts, being without a mouth, have to absorb their amino acids through the body wall; examination of its ultrastructure would make an interesting study of this process. Rediae, on the other hand, are able to ingest host cells and probably obtain nutriment both from the breakdown of such cells and from absorption from the host. The origin of the amino acids concerned in the synthesis of protein in rediae, sporocysts or cercariae has been examined in detail for several species, e.g. *Glypthelmins* spp., *Gorgodera amplicava* and *Echinoparyphium* sp.[37]

The amino acid composition was found to be similar to, in some cases identical with, that of the mollusc tissue (hepatopancreas) or

TABLE 18

Amino acid composition of sera of the lamellibranch Musculium partumeium *and of larvae of* Gorgodera amplicava *parasite in it*
(Data from Cheng[37])

Amino acids	Musculium partumeium Uninfected sera		Infected sera	Gorgodera amplicava Sporocysts		Cercariae	
	Bound	Free	Free	Bound	Free	Bound	Free
Cysteine	+	+	−	+	−	+	−
Arginine	+	+	−	+	−	+	−
Aspartic acid	+	+	−	+	+	+	+
Glycine	−	−	−	−	−	−	−
Threonine	+	+	−	+	+	+	+
Glutamic acid	+	+	−	+	+	+	+
Alanine	+	+	−	+	+	+	+
Tyrosine	+	+	−	+	+	+	+
Valine	+	−	−	+	−	+	−
Tryptophan	+	−	−	+	−	+	−
Isoleucine	+	−	−	+	−	+	−
Leucine	−	−	−	−	−	−	−
Cystine	+	+	−	+	−	+	−
Lysine	+	+	−	+	−	+	−
Serine	+	−	−	+	+	+	+
Asparagine	+	+	−	+	+	+	−
Proline	+	+	−	+	+	+	+

sera, with which it comes in contact (table 18). Moreover, uninfected mollusc sera contains appreciable quantities of free amino acids which are almost entirely missing in the case of infected molluscs. The latter result suggests that the free amino acid pool in the sera of uninfected hosts is being absorbed by the parasites, although these acids could also be used up by the host in the synthesis of new proteins for repair or for defence purposes.

This suggests that the bound amino acids are derived from the hepatopancreas by histolysis of the latter, but some, of course, could have been derived indirectly by transamination (p. 51).   In the case of the sporocysts of *G. amplicava*, the bound amino acid composition bears a striking similarity to the bound amino acid composition of the sera rather than the tissue of the host. This may be explained by the fact that the sporocysts of this species occur within the water tubes (part of the haemocoele) between the inner and outer gill lamellae of the bivalve host which, unlike the hepatopancreas, does not undergo histolysis.

Apparently nothing is known of the intermediate protein metabolism in larval trematodes or its end products.

Lipid metabolism. The lipid metabolism also is little known. The lipid context of sporocyst walls appears to be correlated approximately with the sites which they occupy in the molluscan host.[106] Larval trematodes on the surface of the host intestine or between organs are almost devoid of lipids, whereas those embedded in the host hepatopancreas are rich in fats. There is some relationship between the presence of larval trematodes and the occurrence of large quantities of neutral fats, together with some fatty acid droplets, in the hepatopancreas of molluscs. The parasites presumably stimulate fat synthesis in some way, and, in prolonged infections, fatty acids are released by the breakdown of neutral fats. The sporocyst may transport the simpler fatty acid molecules across its wall and use them in the synthesis of larger molecules of fatty acids.[43]

Effects of Parasitism on Molluscan Hosts

The immunological and host tissue reactions of molluscs against larval trematodes are discussed in Chapter 9. The present section is limited to the major pathological and physiological effects of parasitism on molluscan hosts.

General comments

It is questionable whether any infection of larval trematodes, no matter how light, is not without *some* effect on the mollusc host. The nature and degree of the effects are related to a number of factors, chief of which are: (*a*) the degree of infection, (*b*) the size and age of host, (*c*) the developmental pattern of the larvae (especially with reference to whether a redial stage occurs or not), (*d*) the nature of organs invaded. As is evident from fig. 29, considerable migration may occur within the snail.

Three main types of effect can be distinguished: (*a*) those affecting the hepatopancreas, (*b*) those affecting the gonads, and (*c*) those affecting the general physiology of the host. More than one of these effects can occur in a host at the same time. Cheng and Snyder[42] have given a useful review of the host-parasite relationships between larval trematodes and their hosts.

Effects on foot and hepatopancreas

These range from the slight to the very severe. The most important factor appears to be the number and size of the larvae present. In general, the histological or mechanical damage to the muscular tissue of the head-foot region is not great but the muscle glycogen may be greatly reduced. The early sporocyst stages tend to be spent in this region and presumably such stages do not make such heavy metabolic demands as the later stages. Sporocyst migration appears to take place via blood sinuses and channels rather than up the secretory ducts to the hepatopancreas Although sporocysts may be found in the inter-tubular connective tissue they occur primarily in the haemocoele. The hepatopancreas is usually markedly affected and the number of tubules may be greatly reduced: the glycogen content may decrease, the fat content may increase and the excretory products of the larvae may have a profound lytic effect on the hepatopancreatic cells, causing local histolysis. In some cases, as the number of parasites decreases through release of cercariae, mollusc tissue can gradually return to normal.

A secondary effect of the destruction of the hepatopancreas is the release of pigment from this gland so that the molluscs become pigmented. A typical example of this is seen in the common periwinkle, *Littorina littorea*, whose foot becomes yellow or brown when infected with *Cryptocotyle lingua*[356]; infected specimens are recognizable in this way.

It is clear, then, that the hepatopancreas may be profoundly affected by the presence of larval trematodes. The main changes which have been reported to occur in this organ are listed below:

1. Cells show accumulations of fatty bodies.
2. Vacuoles appear in cell cytoplasm.

3. Karyolysis occurs.
4. Sloughing of tissues occurs.
5. Formation of fibromata (fibrous tumours) and granulomata (granular tumours) occurs.
6. Tunica propria is penetrated by foreign bodies.
7. Amount of glycogen decreases.
8. Histolysis of tissues by excretory products of larvae.
9. Indirect pressure damage from rediae located in gonads.
10. Host cells stimulated to produce granular substances.
11. Direct ingestion of cells by rediae occurs.
12. Release of cell pigments takes place.
13. Cells altered from the columnar to the squamous type.
14. Displacement, but not destruction, of tubules occurs.
15. Abnormal mitosis can occur.
16. The tunica propria may be ruptured.
17. Vessels in the intertubular spaces may be ruptured.

Effect on reproductive system

Molluscan gonads are attacked mainly by those trematodes which possess redial stages in their life cycles. Three types of effect can be produced: (*a*) direct ingestion of gonadal tissue by rediae (parasitic castration), (*b*) sex reversals, (*c*) inhibition of normal gametogenesis.

Parasitic castration. Complete castration of *Littorina littorea* by infections with rediae of *Cercaria himasthla secunda* or *Cercaria lophocerca* has been reported.[257]   In this species, castration is due to ingestion and destruction of the gonads by the rediae. Similarly, the gonads of the limpet *Patella vulgata* may be consumed by invasion by rediae of *Cercaria patellae*.[254]   Castration may not always be permanent, however, and regeneration of gonadal tissue may occur even after severe atrophy.   In some species too, when the gonads are partially invaded, less drastic effects may occur and the external genitalia may be merely reduced in size.

Sex reversal. There are several well documented cases of functionally female snails developing male external genitalia; for instance, females of *Peringia ulvae* infected with larval *Halipegus occidualis* developed small penes.[168]

Inhibition of gametogenesis. A partial inhibition of gameto-genesis can also be caused by partial invasion of a gonad without complete castration resulting.

Effect on the general metabolism

Superficially, trematode infections do not appear to have profound effects on the metabolism of their mollusc hosts but the effects may be greater than is apparent. The glycogen reserves fall, as was mentioned earlier, but the oxygen consumption and fat metabolism appear to be unchanged.[345] In some cases, the presence of infection makes snails more susceptible than uninfected snails to damage from lack of oxygen.[235] Larvae also use free amino acids from the sera (p. 89), so presumably these would not be available for repair of host tissue.

In some cases certain macroscopic effects have been observed, such as thinning and 'ballooning' of shells and gigantism. The latter has been reported in parasitized *Limnaea auriculata*, *Peringia ulvae* and *Littorina peritoides*. Several explanations have been put forward to account for this phenomenon[42]: (*a*) that it results from excessive consumption of food to meet the demands of the parasites; (*b*) that it is an indirect manifestation of parasitic castration, implying that the nutrients normally used for gonad growth are now available for general growth; (*c*) that the parasites supply some growth factor which stimulates tissue growth. A similar phenomenon has been reported for mice infected with the larval stages of the pseudophyllidean cestode *Spirometra mansoides*.[226]

It can be generally concluded, however, that the physiological basis of gigantism in parasitized molluscs remains undetermined. Some workers have also reported that infected snails produce substantially more heat (2·7 times) than uninfected snails—an effect suggesting major metabolic differences.[145]

7 : Biology of the Cercaria

General Organization

General anatomy

A cercaria is basically a juvenile trematode with a tail. In order to understand some of the physiological problems involved in the release of a cercaria from its intermediate host and the pattern of its later stages of development, it is necessary to be familiar with the general anatomy of a cercaria. There is no ' typical ' cercaria; but those most frequently encountered and most widely used in experimental work belong to the Schistosomatidae and the Fasciolidae—the morphology of a typical schistosome cercaria is shown in fig. 31. There are, however, a number of other, different, types of cercaria; these are classified according to several characteristics, in particular the number of flame cells, the type of tail (single or forked), the arrangement and position of suckers and the presence or absence of a stylet. Detailed accounts of cercarial anatomy are given in other texts.[61, 146, 290]

Attention is drawn to the following features. The entire cuticle is covered with minute spines which are particularly concentrated along the anterior fifth of the body and its posterior margin, and also on the acetabulum, where they are arranged in several concentric rings. In addition to spines, the body and tail have several groups of fine, sensory bristles some 10–15 μ long; most of these seem to rise from small cuticular papillae. The digestive system is rudimentary and remarkably uniform, with a mouth situated subterminally in the middle of a powerful oral sucker, an oesophagus and a pair of short intestinal caeca. The nervous system consists of a diffuse, spool-shaped mass of fibres behind the oral sucker, forming a cerebral ganglion from which lead off three pairs of nerves (fig. 32). Many cercariae possess pigmented eye-spots (or ocelli) situated dorso-laterally; occasionally an additional median ocellus is present (fig. 32). In the development of an eye-spot a 'lens'-like structure appears, around which pigment is deposited. The pigment may even spread further and become deposited on the ganglion and longitudinal nerves.[227] The eye-spots are composed of irregularly shaped, coarse brownish or black pigment granules of melanin, a pigment associated with photoreceptors in other invertebrates such as crustaceans and molluscs.

It is interesting to note that, embryologically, an eye-spot develops in a manner similar to that of the eye of a vertebrate (i.e. by an ectodermal cup forming in the vicinity of a nerve cell). Only when the nerve cell occupies a position within the cup do the pigment granules appear. This rather suggests that the production of pigment granules may be the result of an inductive process.

The remnants of eye-spots may occasionally be observed in adults. Tyrosine, which is a precursor of melanin, may be detected in cercariae. The protonephridial system is well developed and connects to an excretory bladder. The flame cell formula is of considerable diagnostic value.

Secretory glands

Types. Cercariae are usually well provided with unicellular secretory

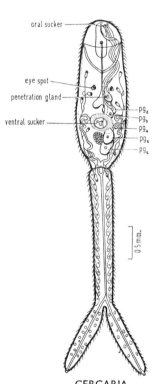

CERCARIA
flame cell = 2[5 + 1]

FIG. 31. Cercaria of *Schistosomatium douthitti* to illustrate the general anatomy of a schistosome cercaria. (after Price, 1931)

glands whose nature and number are related to the later patterns of development. An exception is the cercariae of the Azygiidae and the Bivesiculidae, which are ingested directly (p. 128), and in which glands are lacking. The main developmental patterns are:

> *Type* 1. *Penetrating cercariae*
> (*i*) Those which penetrate the definitive host and develop directly into an adult (e.g. *S. mansoni*).
> (*ii*) Those which penetrate an intermediate host and undergo further development as metacercariae which encyst in some species and remain free in others. The intermediate host is then ingested by the definitive host or in many cases by a further intermediate host (e.g. *Cryptocotyle lingua*).
> *Type* 2. *Encysting cercariae*
> Those which encyst on a substrate, such as vegetation, which is then ingested by the definitive host, e.g. *Fasciola hepatica*. More rarely, cercariae may encyst within the mollusc host. An aberrant cerceria is that of the Haplosplanchidae which becomes a planktonic metacercaria.

Penetration glands. These are best seen in the schistosomes. The number of glands—perhaps best termed in general terms ' *cephalic glands* '—in this type of cercaria, as exemplified by *S. mansoni* or *Schistosomatium douthitti*, has been a matter of some dispute, the point of issue being whether there are five or six pairs. These structures are not easy to distinguish microscopically but it now seems clear that, at least in some schistosome species, there are six pairs, made up as follows: one pair of ' escape ' glands, two pairs of anterior preacetabular (penetration) glands, three pairs of posterior postacetabular (penetration) glands. The confusion has arisen because some authors have not stated whether a description related to a shed cercaria or to one dissected from a snail. The latter is believed to possess ' escape ' glands—a pair of mononuclear gland cells with a faintly basophilic cytoplasm probably related to the escape of a cercaria from a snail; in shed cercariae such glands would naturally be empty. Some authors have failed to detect their presence.[222, 310]

The morphology, staining reactions, histochemistry and chemical composition of the secretions of the remaining cephalic glands have been much studied.[222] Study of the secretions of the glands has been helped by the curious discovery[310] that the post-acetabular glands will discharge themselves if placed in solutions of indian ink. Under these conditions, cercariae deposit a secretion which swells to relatively large droplets of mucoid material. Chemically it is strongly PAS positive and has been shown to consist of mucocomplexes or substances formed by associations of proteins, polysaccharides and lipids. Hydrolysis yields 17·2 per

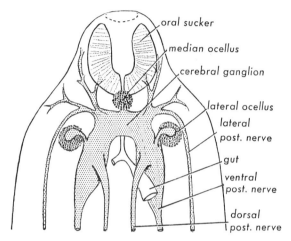

FIG. 32. Central nervous system and ocelli of the cercaria of *Notocotylus ralli*. (after Dönges, 1962)

cent. amino acids, 9·7 per cent amino sugars and an unidentified concentration of steroids.[309] Secretion of the preacetabular, as well as the postacetabular, glands can be stimulated in other ways. Probably the simplest method is to transfer a film of human sebum to a finger (by rubbing the forehead), smear it on a slide and cover the layer with indian ink.

The postacetabular and preacetabular glands have strikingly different staining reactions—purpurin in particular (table 19) stains the preacetabular glands a brilliant red colour when used as a vital stain, whereas the postacetabular glands do not stain.[310] It is believed that the secretion of the postacetabular glands is deposited as the cercariae loop over the skin during exploration, and it seems to have mainly an adhesive function during this period and during the early phases of penetration. In many species, this mucous secretion may also have lubricatory, protective and enzyme-retaining properties. In schistosomes, glands become completely exhausted shortly after penetration. In forms such as *Paragonimus kellicotti*, where the cercaria is of the microcercous type with a very short tail, the discharge of mucus from mucoid cells may play an important part in increasing the possibility of infecting the second intermediate host.[167] Thus, groups of cercariae may become entangled together by mucus and the strands

TABLE 19

Some staining reactions and properties of the glands in schistosome cercariae
(Data from Stirewalt and Kruidenier[310], Rai[250])

Reactors & properties	v = vital stain ' Escape ' glands	a.f. = after fixation Preacetabular glands (2 pairs)	Postacetabular glands (3 pairs)
General description	faintly granular	macrogranular	microgranular
General reaction (a.f.)	basophilic	acidophilic	basophilic
Alizarin (v)	?	pink	not stained
Purpurin (v) (a.f.)	?	red	not stained
Aniline blue (a.f.)	?	not stained	blue
Toluidin blue (a.f.)	?	not stained	blue
Orange G (v) (a.f.)	?	not stained	pale orange
Mallory's triple (a.f.)	faint blue	light red	dark blue
Periodic acid Schiff	?	not stained	purple
Probable functions	secretes proteolytic enzymes?	secretes hyaluronidase, collagenase and mucopolysaccharidase?	secretes mucus

so formed would facilitate the attachment of the cercariae to the appendages of the second intermediate hosts (freshwater crustaceans).

Probably the function of the preacetabular secretion is primarily enzymatic. Enzymatic functions have not been demonstrated in these glands, however, no doubt because of the technical difficulties in obtaining the secretions of individual glands free from those of the other glands. In cercarial suspensions and extracts, however, a thermolabile, hyaluronidase-like enzyme has been identified[307] and also a mucopolysaccharidase and a collagenase-like enzyme.[173, 183, 184] It seems likely that the bulk of these enzymes originate from the preacetabular glands, but some may come from the postacetabular also.

Cystogenous glands. Cercariae which encyst, either on foliage or after penetrating an intermediate host, contain *cystogenous glands* which secrete the multilayered wall of the metacercarial cyst (figs. 33 and 44); penetrating glands may also be present.   In

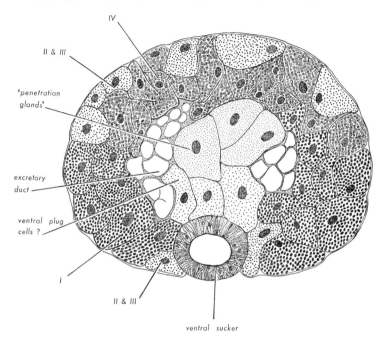

FIG. 33.   Section of encysting cercaria of *F. hepatica* to show the various glands concerned in cyst formation.
The numbers I, II, etc. refer to the cells forming layers of the cyst (see p. 100). See also fig. 44. (after Dixon, 1965)

species such as *F. hepatica* or *Notocotylus ralli*, which do not penetrate into an intermediate host but encyst on vegetation etc., true penetration glands are missing. As the metacercarial cyst has been found to consist of a number of layers, it is not surprising to find that a number of glands are involved.   Some of these have been thought by earlier workers to be ' mucous ' glands; but they clearly form certain layers in the cyst.   In *F. hepatica* for convenience these glands have been simply labelled Group I Gland Cells, etc.[77] (figs. 33 and 44).

Group I Gland Cells. Situated ventrolaterally. (Sometimes termed 'ventral' cystogenous gland cells.) *While the cercaria is still within the snail* these glands secrete a layer outside the body proper.

Groups II and III Gland Cells. Situated dorsally and dorso-laterally. Probably give rise to mucoprotein, acid mucopolysaccharide and neutral mucopolysaccharide layers, which together comprise layers II and III (fig. 44). Sometimes termed 'mucous gland cells'.

Group IV Gland Cells. Sometimes termed 'dorsal batonnet cells'. They are characterized by containing short bacteria-like rods, about $0·75 \, \mu \times 6 \, \mu$. These rods give rise to the keratin layers of the cyst.

Ventral Plug Cells. These are PAS^{+ve} and give rise to the ventral plug region of layer IV (fig. 44).

Ventral Cells. Also PAS^{+ve} and their occurrence around the edge of the central sucker suggests that they may have something to do with attachment to the substrate or with the formation of the thicker central region of layer.

The relation of these various glands to the formation of metacercarial cyst is discussed later (p. 122).

Miscellaneous glands. In addition to penetrating and cystogenous glands, a number of other glands, whose functions are largely unknown, occur in some species of cercaria. Thus, in the

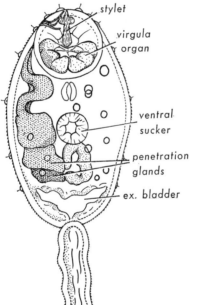

FIG. 34. *Cercaria nyxetica* showing virgula organ. (after Seitner, 1945)

so-called *virgulate* cercaria is found a *virgula organ*—essentially a reservoir containing secretions from one or two pairs of mucoid glands. These and similar mucoid glands, which may be homologous with some of cystogenous glands, develop along the ventral surface under the cuticle of the cercaria as it develops in the snail. The contents of these glands are passed up into the virgula organ (fig. 34). This in turn discharges to the outside, sometimes before the cercaria is released, so that the cercaria is covered in a sticky film of mucus. This undoubtedly has considerable

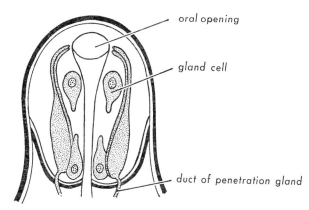

oral opening

gland cell

duct of penetration gland

FIG. 35. Oral sucker of *Alaria arisaemoides* showing presence of gland cells. (after Pearson, 1956)

survival value in enabling the cercaria to become attached to the second intermediate hosts, usually aquatic insect larvae, thus allowing penetration to take place; they may also have protective value against microorganisms. The mucus appears to be made up of acid mucopolysaccharide, lipid and a substance or substances resembling elastin.[236, 237] Hydrolysates of the secretion have yielded arginine, aspartic acid, glycine, glutamic acid, histidine, hydroxyproline, phenylalanine, proline, serine, tyrosine and glucosamine. Alanine, isoleucine, leucine, lysine, threonine and valine were sometimes detected. Tryptophan, sulphur-bearing amino acids and galactosamine were not detected.[236, 237]

Some cercariae (e.g. *Alaria arisaemoides* and cyathocotylids) and some mesocercariae (e.g. *A. arisaemoides*) have been shown to possess unicellular glands, of unknown function, within the oral

sucker (fig. 35).[241] They are also developed in the tetracotyle of
S. elegans.[242] Some cercariae possess glands, in association with

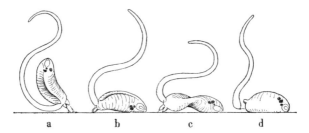

FIG. 36. Behaviour of cercaria of *Notocotylus ralli* during
early stages of encystment. (after Dönges, 1962)

the posterior locomotory processes (fig. 36), which apparently
secrete an adhesive material.

Emergence of Cercariae from Molluscan Hosts

Factors influencing emergence

As might be expected, many physico-chemical factors in the
environment affect the release of cercariae from molluscan hosts—
temperature, light, humidity, oxygen and pH. The mechanism

TABLE 20

Emergence of cercariae of F. hepatica *from* L. truncatula *under different
conditions*
(Modified from Kendall and McCullough[159])

Conditions	% snails from which cercariae emerged
Snails undisturbed	2
Snails returned to same water after mechanical disturbance	4
Snails in water previously occupied by other snails	12
Snails placed in fresh pond water	54

of emergence is not straightforward, however, for simply changing
the water will often not only induce emergence of cercariae but
also increase the activity of the snail itself. A general stimulation
of a resting snail may thus be the principal factor in causing

emergence in many species. This is well shown in *F. hepatica* (table 20) where rain water has a similar stimulating effect.

Light. In the case of *S. mansoni*, possibly correlated with the habits of the definitive hosts, shedding of cercariae takes place between 9 a.m. and 2 p.m. in direct sunlight. In *S. bovis*, cercariae will emerge regularly at all hours of the day and night although more are shed in strong light than in diffused light or in total darkness.[180]   Some species of cercariae (e.g. *Cotylophoron cotylophorum*) do not emerge at all in the dark.[332]   In *Schistosomatium douthitti*, on the other hand, shedding takes place only in the evening or at night, as is clearly shown from the data in fig. 37. This also illustrates how the emergence pattern can be reversed by keeping snails in the dark during the normal daylight hours.

Temperature. Temperature can be expected to have a ' threshold effect ', i.e. below and above certain temperatures emergence will not occur. In *F. hepatica*, no emergence will occur below about 10° C. The effect of high temperatures (above about 26° C.) cannot be examined in the case of *F. hepatica* on account of their

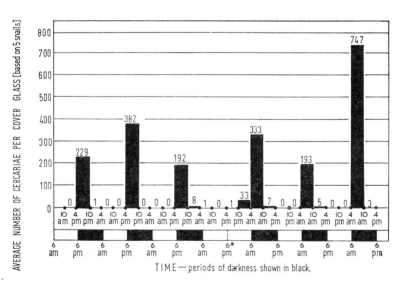

FIG. 37.   Effect of light on the emergence of cercariae of *Schistosomatium douthitti* from *Lymnaea*. (after Olivier, 1951)

effects on the snail; emergence can, however, still take place at this temperature.

In the case of *C. cotylophorum*, cercariae will not emerge below 24° C. or above 35° C.; in this species, the maximum emergence appears to occur between 30° C. and 35° C.

pH. Mass emergence of cercariae appears to be limited to a *p*H range of about 6·5–9·5, but a little emergence will take place just beyond these limits. Below *p*H 3 and above *p*H 11, no emergence takes place.[332]

Process of emergence

There appear to be four main methods whereby cercariae escape from the tissues of the mollusc host:

(*i*) *active* escape from a blood-vessel through fixed openings in the integument (e.g. *Alaria canis, A. arisaemoides, Strigea elegans, Neodiplostomum intermedium, Apatemon* sp., some schistosomes and some echinostomes).[241, 242, 243]

(*ii*) *active* escape through the intact integument (e.g. *S. mansoni*[84]).

(*iii*) *passive* extrusion of masses of cercariae (e.g. *F. hepatica*[159]).

(*iv*) *active* escape of daughter sporocysts, containing cercariae, which are then eaten by the intermediate host (e.g. *Dicrocoeliodes petiolatum*).

' Active ' escape implies that cercariae actively move through the snail before emerging through an ' escape ' pore. In most cases, the main movement is through the snail blood vessels. In *S. mansoni*, cercariae leave the sporocyst in the perihepatic (blood) space and travel down the haemocoelic spaces and veins of the snail; the majority arrive via the perirectal spaces at the pseudo-branch and collar, at which sites they penetrate the integument and emerge to the exterior.[84] Histolytic secretions—possibly produced by the ' escape glands ' (p. 96)—may assist this process in some species.

' Passive ' escape, as exemplified by *F. hepatica*, implies that cercariae make no active movements to escape but are forced through the integument. In this species, cercariae congregate in the perivisceral space surrounding the distal part of the gut of the snail. Shortly before emergence, an area contiguous with the anus becomes tumid and teat-like. When the pneumostome closes, cercariae begin to be extruded through the teat-like process.[159]

It seems likely that in most cases emergence of cercariae is a complex process and that activity on the part of *both* the cercariae *and* the snail are required to bring it about. The points of emergence from a particular snail seem to be fixed; the first cercariae break through at one or more points and all subsequent cercariae escape through these.[241]

The process of escape of sporocysts containing cercariae, in a number of dicrocoeliid genera, is a remarkable process, which has been described in some detail for *D. petiolatum*. In this species, the secondary sporocyst develops a thick internal wall or *endocyst* which renders it somewhat resistant to desiccation. Sporocysts pass from the visceral mass into the pallial cavity and are extruded at regular intervals through the pneumostome as white, lemon-shaped cysts. Emission can occur during the night or day but is influenced by humidity. Emerged sporocysts (which are normally ingested by the terrestrial isopods *Armadillidium vulgare*, or *A. officinalis*) can survive for at least twenty-four hours in relatively humid conditions. The chemical nature of the endocyst has not been determined, but is believed to be collagen.

Behaviour

The behaviour of cercariae, which is an important aspect of trematode ecology, has been little studied. As is shown below, the responses to light, gravity, agitation and touch, appear to play a vital rôle in increasing the probability of infection of the definitive or intermediate host. The degree of importance of these stimuli will depend largely on the nature of the biotope and the behaviour of the host.

Locomotion

General movements. Most cercariae are free-swimming but some (e.g. *Paragonimus westermani*) have very reduced tails and crawl on the substrate surface with a leech-like movement. Four general types of cercarial movement can be recognized:

(*i*) the ordinary swimming movement in which, in the case of furcocercaria, the cercariae generally swim with the tail thrust forward. The locomotor power is furnished by the tail, the contraction and extension of the bipolar muscle cells producing a left and right vibration. Analysis of the movement from electronic flash photomicrographs shows that the wide sides of the furca make such an angle with the water that they function as a propeller (plate III, fig. B). Schistosome cercariae can, however, swim either head-first or tail-first depending on the angle between

TABLE 21

Comparative behaviour of some schistosome cercariae
(Data from Lee[175])

S =surface film ; W =wall of container ; N.A. =not attached

Species	Definitive host	Phototaxis	Geotaxis	Attached to
Trichobilharzia ocellata	Bird	+ve	−ve	W
T. physellae	Bird	+ve	±ve	W
T. stagnicolae	Bird	+ve	?	N.A.
Gigantobilharzia gyrauli	Bird	o?	−ve	S
G. huronensis	Bird	?	−ve	S
G. huttoni	Bird	+ve	−ve	S
Austrobilharzia variglandis	Bird	?	−ve	S
A. penneri	Bird	o	−ve	S
Heterobilharzia americana	Mammal	+ve	−ve	N.A.
Schistosomatium douthitti	Mammal	o	−ve	S
Cercaria littorinalinae	Bird	+ve	−ve	S
Cercaria elongata	Unknown	+ve	−ve	N.A.
Cercaria tuckerensis	Unknown	?	+ve	(Crawls)

the furca.[244] The movements of simple-tailed cercariae lacking furca (e.g. *F. hepatica*) have not been examined in detail; in echinostome and plagiorchid cercariae the tail is thrust forward while swimming.

(*ii*) a vibratory movement—characteristically seen in schistosomes—in which the ventral sucker attached to the surface film is used as a pivot;

(*iii*) a sinking movement, in which the cercaria sinks through the water with the entire body elongate but inactive;

(*iv*) a looping movement, sometimes referred to as a ' measuring worm ' movement, which is performed at the surface film or at the bottom.

In the resting phase, free-swimming cercariae often take up

characteristic positions which can be of diagnostic value (table 21). They may hang head end downwards with the furca extended at an angle of 90°–120° to each other, sometimes with the head bent upwards (fig. 38); they may attach to the surface film with the ventral side up and tail bent dorsally; they may be suspended below the surface, tail up, using just enough energy to maintain their position; or they may sink to the bottom with furca curled, ascending and descending as stimulated (see below). The behaviour pattern varies from species to species and, presumably, with the extent of the energy reserves remaining.

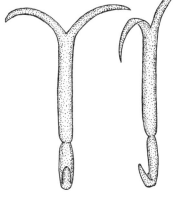

FIG. 38.   Furcocercariae in typical resting positions. (after Dönges 1964)

Spontaneous swimming activity. Many species of free-swimming cercariae show intermittent swimming activity, so that at any one time only a small percentage of a cercarial swarm will be actually swimming. This phenomenon has been analyzed in detail for *Posthodiplostomum cuticola*.[82, 83] In this species, at 20° C. and a light intensity of 500 lux, swimming movements of about 0·2–0·8 sec. occur every 25 sec. The proportion of cercariae showing this spontaneous activity varies logarithmetically with the intensity of the light within a temperature range of 13·5°–22° C. In complete darkness the spontaneous activity is about 1 per cent.; at 300 lux it is 2 per cent.; at 2000 lux it rises to 4 per cent.; at 3000 lux (or sunlight) it is about 5 per cent.

Phototaxis

Many species of cercariae exhibit positive phototaxis (table 21), but the phenomenon has not been much studied. Positive responses are presumably restricted to cercariae which possess eye-spots (see p. 95). An apparently simple phototactic response may

hide a complex pattern of responses, as is shown in the case of
P. cuticola. In this species as shown below, *change* in intensity—
such as might arise from the shadow of a fish—serves as an
important stimulus affecting the general pattern of spontaneous
activity which occurs in a cercarial swarm. The following experiments were carried out with *P. cuticola*: [82, 83]

Change in light intensity

1. *Sudden increase.* Cercariae exposed to a constant source of
1000 lux became adapted to this and had a period of spontaneous
activity $\geq$ 3 sec. When the intensity was raised to 2000 lux, after
a space of 3–6 sec. renewed activity commenced, and at 7–10 sec.
about 20 per cent. of the cercariae showed activity, only to become
inactive again by 25 sec. (fig. 39A). Stepwise increase in intensity
thus results in (*i*) an immediately effective inhibition of not less
than 3 sec. duration, and (*ii*) a related period of post-inhibitory
activity.

2. *Sudden decrease.* Reduction in the light intensity produced a
very rapid and widespread response. Thus, in a cercarial swarm
adapted to 500 lux, almost 100 per cent. of the organisms responded
to a sudden reduction to 100 lux or less. When the intensity was
reduced from 2000 to 1000 the response was only 80 per cent. (fig.
39B). The response is the same even though the light reduction is
of very short duration $\geq$ 0·2 sec. (fig. 39C). This response can be
termed a ' shadow ' response and has been described for a number
of species.

3. *Change of angle of illumination.* That the ocelli are concerned
in detecting *the direction* of illumination can readily be demonstrated. Thus if this was turned through 180°, some 50 per cent.
of the cercariae become active immediately. This is interpreted
as a response by those cercariae whose ocelli-containing side
experienced the shadowing effect.

4. *Alternating increase and decrease.* If the light intensity was
increased from 1000 lux to 2000 lux for 0·3 sec. then reduced to
0 for the same period and then back to the original 1000 lux,
spontaneous activity ceased for about 3 sec. (fig. 39D). The extent
to which this inhibition depends on the time increment $\triangle t$

between the increase and decrease is shown in fig. 39E.   For example, when $\triangle t=6$ sec., 30 per cent. of the cercariae react.

5. *Periodic decrease and increase.*   If the intensity was decreased and increased rapidly for 0·5 sec. at a time, cercariae responded to the first three stimuli only;   thereafter adaptation occurred (fig. 39F).

6. *Linear variation.*   When the light was reduced linearly, 2000 to 0 lux in 2 sec., and then increased at the same rate (solid line fig. 39G), a complete inhibition of spontaneous activity

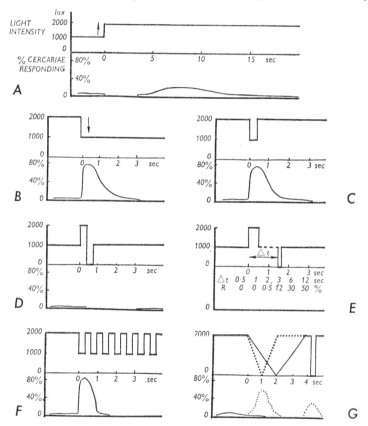

FIG. 39.   Reactions of the cercariae of *Posthodiplostomum cuticola* to optical stimuli.   Temperature 21–26° C.   For explanation see text, p. 108. (after Dönges, 1964)

occurred. However, this inhibition was carried over to a subsequent light reduction challenge (1000 to 0 lux in 0·3 sec.). When, however, the intensity gradient was steeper (1000 to 0 lux in 1 sec.), a partial activity response was restored (dotted line fig. 39G).

Response to water turbulence

If the water in which cercariae are contained is agitated by stirring, cercariae of *P. cuticola* show 100 per cent. activity. The shadowing reaction is not concerned, since in *P. cuticola* the reaction is equally strong in weak light (< 10 lux)—i.e. at an intensity when a shadow response does not occur. The sensory elements concerned may be the filaments on the body and tail stem. Experiments have shown that the current response is generally stronger than the shadow response.

Thigmotaxis

If a cercaria of *P. cuticola* swims against a stationary fish, it usually swims away. If the fish is moving, however, the cercaria as a rule responds rapidly, becomes agitated and will attempt to penetrate.

Chemotaxis

There is no evidence that chemotaxis is involved in cercarial location of a host.

Analysis of responses

The foregoing results show that swimming activity may be induced either by negative changes in light intensity or by mechanical stimuli. The latter produce only activating stimuli, which, it can be shown experimentally, are not affected by simultaneous optical stimuli. This disproves any connection in the central nervous system between optical and mechanical reception. Negative changes of light intensity ($-\triangle lx$, fig. 40, i.e. shadows) stimulate the optical receptor organs (the ocelli) to transmit *activating* impulses to the C.N.S. and positive changes ($+\triangle lx$) stimulate the transmission of *inhibitory* impulses. The spontaneous activity displayed by cercariae may be incited either by a pace-maker of the C.N.S. or by the optical system through continuous discharge of the receptor organs.   Illumination of uniform intensity activates the pace-maker mechanism, with the result that spontaneous activity is increased; in contrast, stepwise positive changes of intensity inhibit it.[83]

Biological significance of phototactic and turbulent responses

The analysis of responses outlined above refers to the cercariae of *P. cuticola*, and their importance to the organism can be

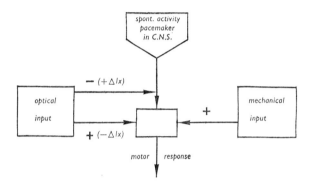

FIG. 40.   Diagram showing the functional relationship between motor responses to optical and mechanical stimuli in the cercaria of *Posthodiplostomum cuticola*. (after Dönges, 1963)

viewed against the life cycle and biotopes of both host and parasite of this species only.[83]   Analysis of other species may reveal somewhat different patterns of responses.

Under relatively stationary lighting conditions, the shadow response would stimulate cercarial activity if caused by a fish, resulting in greater likelihood of cercaria and fish making contact. Likewise, the turbulence caused by the swimming or fin movements or the respiratory currents of fish would induce cercarial activity and increase the probability of infection taking place.   On the other hand, the periodic shadowing by the reeds in the natural habitat would stimulate no great activity, nor would the shadowing produced by clouds.   Thus the glycogen reserves of the cercariae would not be wasted by biologically unimportant stimuli.

Geotaxis

Most species of cercariae exhibit negative geotaxis (table 21), a phenomenon studied in detail in the cercariae of *Opisthioglyphe ranae*.[313]   Cercariae freshly released from the snail *Lymnaea stagnalis* migrate rapidly to the surface if placed in a suitable container such as the tube shown in fig. 41.   This is presumably the result of the spontaneous swimming activity described above. The measured velocity is about 1·06 mm./sec., or approximately

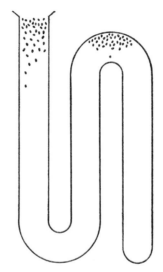

FIG. 41.   Distribution of cercariae of *Opisthioglyphe ranae* showing the negative geotropism of these larvae. (after Styczyńska-Jurewicz, 1961)

380 cm./hr.   The migration takes place as rapidly in complete darkness as in an illuminated tube, which proves it to be independent of light intensity.

Cercariae which reach the surface continue to swim to the film and back a short distance from it, but generally stay just below the surface.   The time of swimming beneath the surface is closely related to the length of ' free life ' of the cercaria which has already elapsed.   It can generally be said that cercariae show geotactic responses for about 4–8 hours but can live much longer in a vessel if the depth of the water in it is small.   Thus, in a water column of

1·5 cm. the cercariae still showed a geotactic response after 70 hours.

After a period at the surface, cercariae fall passively to the bottom. A few may do this within an hour or so (fig. 42) but may rise again after a further period. Eventually, however, all cercariae fall to the bottom. Under any normal form of stimulation, such as shaking the container, they are again activated to show a geotactic reaction and rise up in a cloud. After a few moments,

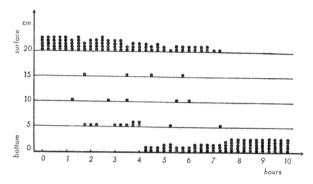

Fig. 42. Vertical distribution of five specimens of *Opisthioglyphae ranae* cercariae observed over a 10-hour period at 15-minute intervals. (after Styczyńska-Jurewicz, 1961)

however, the majority fall passively to the bottom once again. Even at the bottom they will swim spasmodically for a time until, finally becoming exhausted, they lie on the bottom and die.[313]

Since the life of a cercaria is so limited, it is clearly important that the responses of the cercaria shall be as closely related as possible in a biologically advantageous way to those of the intermediate host, so that the maximum chance of contact may result. A curious form of negative geotaxis is seen in the non-swimming cercariae of *Asymphylodora amnicolae*. When these emerge from their snail host they climb to the top of the snail and cling to the tentacles. This behaviour pattern appears to be related to ease of infection of the second intermediate host. This is an uninfected species of the same snail; they make contact with it, penetrate it and encyst, eventually becoming sexually mature as progenetic forms.

T.—H

Viability

Some factors affecting viability have already been discussed above. The life period of an individual cercaria is very much dependent on the extent of the glycogen reserves, and any environmental factors such as temperature or turbulence, which may stimulate expenditure of energy, will clearly lower viability. In the case of *S. mansoni*, it has been shown that the rate of flow of a stream in which the cercariae are carried is an important factor in determining infectivity, and when the water velocity exceeds 2·7 miles per hour the potential infectivity of *S. mansoni* cercariae is reduced.[249] The distance a cercaria is carried also affects its infectivity; and in field experiments it has been shown that the greater the distance away from the original source of release, the less percentage of infection results.[249] Presumably some cercariae have less energy reserves than others, and if they are swimming any distance, even with the current, the food reserves may fall below that necessary for the (undoubtedly energetic) process of penetration.

General Physiology

General metabolism

The cercaria is a short-lived non-feeding stage in the digenetic trematode life cycle, and it is not surprising to find that cercariae are rich in glycogen; the metacercaria of *Gynaecotyla adunca*, for example, contains 2·7 μg. glycogen/mm.[3]. The body proper is the only portion which continues to develop in the next host (intermediate or definitive), and there is relatively more glycogen in the body than in the tail.[36] The glycogen distribution in the larval stages of a number of species of trematodes has been studied.[105, 107]

Although the lipid metabolism does not appear to be an important source of energy in the intramolluscan cercariae,[40] it can be utilized as such in the free-living stages of certain species which have active swimming cercariae.[106]

The period of survival of a cercaria depends largely on the extent of its food reserves in relation to its body size and its general metabolic activity. With regard to size, a wide range is found. For example, the cercaria of *Himasthla quissetensis* is approximately 100 times larger than that of *G. adunca*.[338] Although a size range is usually found with each species, this is probably related to contractibility during fixation rather than to variation in weight of cercaria. Thus, ten cercariae of *H. quissetensis* were found to have a nitrogen content of 0·24–0·48 μgm. N/worm. No growth, of course, occurs during the cercarial stage.

At temperatures of 20°–24° C. most species of cercariae probably exhaust their glycogen reserves within 8–12 hours. The addition of glucose to the water can, however, greatly increase the survival period. Thus, cercariae of *Cotylurus brevis* die in 8–10 hours at 23°–24° C. in water, but live for 30 hours or more in 0·3 per cent. glucose.[108] A curious point is that cercariae are unable to resynthesize glycogen from glucose; nor does the latter stimulate the respiration (table 17).[341] A cercaria is also unable to utilize succinate or proline.[341] It appears then that, during the cercarial phase, a larva can utilize substrates for energy purposes but not for synthetic purposes. There is clear evidence (p. 153) that in some, if not all, species a ' trigger ' mechanism of some kind is involved and that this is responsible for initiating the next phase of development, which is concerned with both synthetic and energetic processes.

Little is known regarding the metabolism, carbohydrate, lipid or fat metabolism of cercariae; and whether or not the Embden-Meyerhof pathway and the Krebs cycle are followed has not yet been investigated.

Respiration

Influence of O_2 tension. Since cercariae are free-living, a predominantly aerobic metabolism can be expected. This is borne out by the fact that, in contrast to the adults, they generally cannot survive for prolonged periods under anaerobic conditions. Thus the cercariae of *Zoogonus rubellus*, which are tail-less, relatively inactive forms,[144] will survive 30–36 hours in normal aerobic conditions, but after 12 hours under completely anaerobic conditions 75 per cent. die. Under anaerobiosis, movement was found to cease within 9 hours but some cercariae recovered on exposure to air within 30 seconds. The cercaria of *Himasthla quissetensis*, on the other hand, is a more active form and dies within 6–8 hours without oxygen. Shed cercariae of *S. mansoni*

die even more quickly from oxygen lack, becoming inactive after
1 hour; 75 per cent. die 4 hours after emergence; cercariae within
a snail, however, can withstand 6–16 hours without oxygen.[235]
Studies on the influence of oxygen tension have been confined to
marine forms. The oxygen tension of sea water at summer temperatures has been found to range from 1·6 ml./l at night at low
tide, to 11·0 ml./l at low tide in bright sun, the latter high figure
being related to photosynthesis during daylight.[336] Thus, during a
24 hours' period cercariae are exposed to a wide range of oxygen

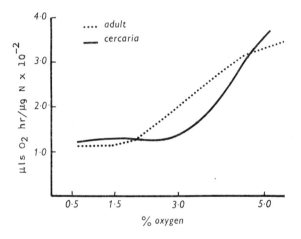

FIG. 43. Oxygen uptake of cercariae of *Zoogonus
rubellus* (solid line) and adult of *Gynaecotyla adunca*
(dotted line). (modified from Vernberg, 1963)

tensions. Other marine organisms which live in such a fluctuating
environment appear to be able to survive periods of anaerobiosis,
but oxygen is ultimately necessary; and this seems to be the
situation with marine cercariae.

The cercaria of *H. quissetensis* behaves as a typical conformer
and the respiration rate decreases in direct proportion to the decrease
in oxygen tension. That of *Z. rubellus* on the other hand is a
conformer only to between the 5 per cent. and 3 per cent. oxygen
levels; below 3 per cent. and above 5 per cent. the oxygen consumption remains relatively uniform (fig. 43 and table 22). After
10–12 hours no oxygen debt is apparently accumulated, since the

oxygen consumption appears to be the same as in organisms which had oxygen available (table 22); but technical difficulties with regard to speed of estimation make this result open to question.[144]

Influence of temperature. It has been shown (p. 31) that the metabolic response of adult trematodes can be correlated fairly closely to the temperature encountered by each species in the definitive host. It is interesting to find that this metabolic response can be detected early in the life history in both the sporocyst and the cercarial stage in some cases.

TABLE 22

Oxygen consumption of the cercariae of Zoogonus rubellus,* *at extremes of oxygen tension*

(Data from Hunter and Vernberg[144])

% Oxygen	No. of worms	$\mu l/O_2/hr./mm.^3$	
		Range	Average
5	9	2·98–8·17	4·52 ±0·46
21 (air)	19	3·17–9·50	5·35 ±0·40
100	9	4·94–9·45	6·66 ±0·56
After 10–12 hours anaerobiosis	10	3·40–8·61	5·18 ±0·62

* Originally identified as *Gynaecotyla adunca.*

Thus the cercariae of *Z. rubellus*, which utilize a fish definitive host, die within half an hour at 39° C.—a temperature about 9° C. above the maximum to which it would normally be subjected in nature.[334] In contrast, the cercariae of *H. quissetensis*, which utilize a bird host (body temperature 40°–41° C.) can withstand 6 hours at 41° C. This interesting result suggests that physiological pre-adaptation, which has enabled the parasite to utilize a warm blooded animal as a definitive host, has occurred in this genus.

A further probable indication of pre-adaptation is seen for the Q_{10} in the case of *H. quissetensis*. It has generally been found that in free-living animals the Q_{10} is low (1·5–2) within the *normal* temperature range of its habitat and it is interesting to note that the Q_{10} of the cercariae of *H. quissetensis* between 30 and 40° C. is 1·5—a figure very similar to that for *adults* of other trematode species, such as *Gynaecotyla adunca.*[340]

Relationship of body size to O_2 uptake. It has been generally found that in free-living invertebrates there is a significant relationship between body size and oxygen consumption—the larger the organism the less the O_2 uptake per unit mass.[365] The same has been found to hold approximately for cercariae. In such small organisms body weight is most conveniently measured in terms of total nitrogen, for the estimation of which some excellent micromethods are available.[18] In trematodes, size appears to be a more reliable indicator of oxygen uptake than activity. Thus, although the echinostome cercaria of *H. quissetensis* is about four times the size of the tail-less cercaria of *Zoogonus* sp., and much more active, the oxygen consumption of the latter is about twice that of the echinostome.[338]

8 : Development within Definitive Host

Fate of Cercaria Prior to Invasion

A released cercaria may reach its definitive host in a number of ways:

(*a*) It may encyst directly on vegetation or a similar substrate, which is ingested by the definitive host.

(*b*) It may penetrate and encyst in a second intermediate host which normally forms part of the natural diet of the definitive host. (More rarely, a third intermediate host is involved.)

(*c*) It may directly penetrate the skin of the definitive host.

(*d*) It may be eaten directly by the definitive host (e.g. Azygiidae).

Direct encystment

General process. The classical example of a cercaria encysting directly on vegetation is that of *F. hepatica*, whose cercaria contains cystogenous, but not penetration, glands. The process of encystment, which can be conveniently carried out on a microscope slide, takes place extremely rapidly and is completed within about an hour. Briefly, it appears to take place as follows.

A cercaria comes to rest, attaches itself by the ventral sucker and becomes flattened. It then adheres at the periphery and the body contracts inwards releasing the outer layer of the cyst which has been already preformed (see p. 122).[78] Simultaneously, as the embryonic 'epithelium' is shed and the outer layer laid down, the tail breaks and becomes rigid, usually remaining attached to the outer cyst wall.

In *F. hepatica*, the tail becomes separated from the body early in encystment but usually remains attached to the body during the process and continues to lash energetically from side to side until a vigorous movement finally detaches it.

The snowy-white covering of the cyst slowly hardens and darkens, due probably to quinone-tanning (see below), and the process is complete. The organism lies orientated within a cyst so that its ventral sucker is, in fact, ventral.

Some species (e.g. *Notocotylus ralli*), make use of adhesive secretions from posterior processes to become attached to the substrate before encysting[81] (fig. 36). The cercaria holds on to the substrate with its locomotory processes and releases a stringy, sticky secretion, often turning backwards while doing so. The body next presses itself on to the substrate by its ventral surface and secretes the base of the cyst on to the substrate. It then twists over so that the ventral side is uppermost and the rest of the cyst is secreted.

The process of secretion of a cyst is more complex than is apparent from a macroscopic description, and in order to understand it more fully, the microscopic structure of the wall must be examined. The cyst of *F. hepatica* is considered below:

Structure of cyst wall. The structure of the cyst wall as revealed by light and electron microscopy is complex.[77, 79] It consists essentially of an *outer cyst* which can be readily dissected off with needles from an *inner cyst*. These are made up of the following four layers:

Outer cyst ⎰ I. Tanned protein layer
⎱ II. Fibrous layer
Inner cyst ⎰ III. Mucopolysaccharide layer
⎱ IV. Laminated (keratinized) layer

The structure of the cyst (fig. 44) is not constant throughout and in particular some layers are absent or reduced in the ventral region where it is attached to the substratum. The outermost tanned protein or sclerotin layer (I) is white when formed and gradually, over a period of several days, becomes dark brown and unreactive. The enzyme involved in quinone tanning systems (see p. 64)—phenolase—has been detected histochemically in the cercaria of *Fascioloides magna*,[31] and the other two components, basic proteins and phenols, in the cyst of *F. hepatica*.[77] The fibrous layer (II) appears to be made up of strands of acid mucopoly-saccharide mixed with strands of mucoprotein. The mucopoly-saccharide layer (III) is made up probably of three sub-layers—an outer, muco- (or glyco-) protein region (IIIa), a middle, acid mucopolysaccharide region (IIIb), and an inner, neutral mucopoly-saccharide region (IIIc). The laminated layer (IV), which is one

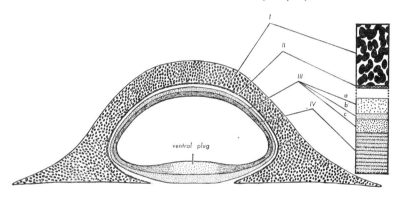

FIG. 44.   Structure of metacercarial cyst of *Fasciola hepatica*. The numbers I, II, etc. refer to the various layers of the cyst (see p. 120). See also fig. 33. (after Dixon, 1965)

of the major layers, is made up of sheets of a keratin-type of protein in an amorphous matrix with lipid and protein components. Under the light microscope this layer is homogeneous except for the thickened central ventral region which forms a 'ventral plug' and is composed of neutral mucopolysaccharide. It is interesting that the ventral plug is made of mucopolysaccharide and not keratin. This may be related to the ability of the parasite to synthesize a mucopolysaccharidase, the existence of which is known in trematodes.[77] If the cyst was completely enclosed in keratin, a keratinase would be necessary to effect escape; it is known that the ability to synthesize this enzyme is rare in the animal kingdom.[77]

Cyst formation. Cyst formation takes place rapidly, and it is not easy to determine how the various layers are formed. The contribution of the various types of cystogenous cells (fig. 33) in *F. hepatica* is as follows:[77, 79]

Group I cells. These cells secrete a layer outside the cercaria *while it is still within the snail*; this layer rapidly becomes brown and becomes the outermost tanned protein layer of the completed cyst. Electron microscope studies have shown that this layer is still contained within a cell membrane although the secretion is outside the body proper. This accounts for the view of some workers that during cyst formation a ' primitive epithelium ' is sloughed off.

Group II and III cells. These secrete the mucopolysaccharide layers which comprise layers II and III. These layers are not secreted until after the cercaria has attached itself to the substrate. The central region of these cells stains somewhat differently from the peripheral cells.

Group IV cells. These cells contain rods or batonettes which are easily visible under light microscopy. Electron microscope studies show that these form the keratin layer (IV). The rods occur in bundles, each made up of some hundreds of rods, lying parallel to each other. Each rod forms a tightly rolled continuous sheet, similar to a roll of carpet, and these sheets unroll at the surface of the organism to form sheets of the keratin layer (plate II, fig. D). Each rod measures about $0.75 \mu \times 6 \mu$ and when unrolled gives a sheet of about $6 \mu \times 30 \mu$.

The formation and chemical nature of the metacercarial cyst of *Notocotylus urbanensis*, which has five layers, has also been studied.[130, 285] Unlike *Fasciola*, the cyst walls of this species contain little mucopolysaccharide and are mainly muco- and/or glyco-protein. The outer layer is also only a temporary structure and is destroyed by bacterial decay after a few days. The remaining layers are extremely resistant. Chemically the cyst walls contain little carbohydrate, and no collagen or chitin; the cyst is thus almost entirely protein. Collagen has been found, however, in the cyst wall of the metacercaria of *Ascocotyle*.[181] In *Posthodiplostomum minimum*, the single-layer wall is made up of carbohydrate-protein complexes.[14, 198]

Encysted metacercariae are usually stated to be ' very ' resistant to environmental conditions; but few precisely controlled experiments have been carried out. The cysts of *F.*

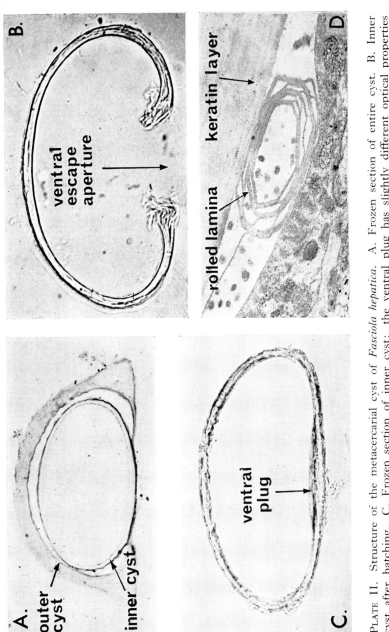

PLATE II. Structure of the metacercarial cyst of *Fasciola hepatica*. A. Frozen section of entire cyst. B. Inner cyst after hatching. C. Frozen section of inner cyst; the ventral plug has slightly different optical properties from the rest of the cyst. D. Electron micrograph of the formation of layer IV (see Fig. 44 and p. 122), the keratin layer. The metacercaria secretes cylindrical rods like laminated, tightly rolled stair carpets. A partly rolled rod can be seen at the surface of the metacercaria laying down another lamina on the inner surface of the keratin layer. × 40,000. (after Dixon, 1965; Dixon and Mercer, 1964)

hepatica, for example, have been reported to be viable after over-wintering in Russia.

Encystment after penetration

Many species of cercariae penetrate a second intermediate host before encysting. The metacercariae of *Clonorchis sinensis*, for example, penetrate various species of cyprinids. Such forms possess both cystogenous *and* penetration glands. Very little is known about the process of encystment in such species, or of the chemical nature of the cyst. The presence of a metacercarial cyst usually invokes a tissue reaction on the part of the host and the cyst becomes encased in host fibrous tissue forming a capsule, so that the cyst is of both parasite and host origin. Some workers distinguish the host portion as ' capsule ' and the parasite portion as ' cyst '.[274] The term ' cyst ' however, is almost universally used in a more general sense. In *Uvulifer ambloplitis*, a thin layered hyaline wall is secreted by the parasite and this is covered by fibroblastic tissue from the host.[142] Black pigmented cells resembling melanophores often surround the fibroblasts, the whole forming the familiar ' black spot ' of fishes (plate III, fig. A, facing p. 168). Yellow spots are formed in some instances. The nature of these tissue reactions is discussed in Chapter 9.

Metacercarial cysts have been reported from a wide range of both vertebrate and invertebrate hosts; fish, arthropods and molluscs are particularly involved. Metacercariae are found also in planarians and leeches. A number of useful surveys of these intermediate hosts have been made.[125, 151, 274, 323]

It is interesting to note that in some cases encystment does not occur immediately after penetration into insect larvae but is delayed until pupation takes place. This is the position with the bat fluke *Prosthodendrium chilostomum* which penetrates into caddis fly larvae and remains unencysted until pupation takes place.[17] This synchronization of the host and parasite, which appears to have an endocrine basis in some cases, may have some survival value and may be related to the ecology of the host. The rôle of some of the glands present in cercariae, which penetrate and encyst, has not been determined. Thus, the contents of the virgula organ (fig. 34) may persist after penetration and encystment, and do not appear to play a part in either of these processes.[276] Encysted metacercariae appear to be able to absorb nutrients and grow and differentiate the anlagen of the genitalia. In some cases metacercariae are markedly progenetic and maturation is achieved within the cyst. Progenesis is discussed further on p. 139.

Penetration without encystment

Into an intermediate host. The cercariae of many species of strigeids penetrate the intermediate host and develop as metacercariae without encystment. These may penetrate sites such as muscles, where a strong host reaction may occur. Alternatively, they may enter sites such as the lens, in which host tissue responses do not occur, or sites such as nervous tissue, in which they are not strongly developed. Such unencysted metacercariae are known as ' diplostomulum ' or ' tyrodelphys '. More rarely, e.g. in *Alaria* spp., a third intermediate host is involved, and the cercaria develops into a mesocercaria—a prolonged developmental stage which possesses cercarial, not metacercarial, features.[241] The mesocercaria becomes a metacercaria in the next host. Useful metacercariae for physiological studies are those of *Diplostomum phoxini* (brain of minnow) and various species in the lenses of fish, e.g. *Diplostomum flexicaudum*. Detailed study of the penetration and migration of one of the last mentioned species, *Cercaria X*, has been made and serves as a useful model in discussing the physiology of this type of development.[86]

(*i*) *Method of penetration.* On coming into contact with the host tissue, a cercaria attaches itself firmly by the ventral sucker to the surface of the fish. This particular species of cercaria is unusual in also possessing rows of hooks which provide additional anchorage. After attachment, movement is leech-like—consisting of alternative attachment by oral and ventral suckers; the larva appears to ' test ' the substratum while it moves. The tail apparently plays no part in this movement. As the oral sucker presses against the host surface, droplets are extruded from the ' penetration ' glands (p. 96) and eventually penetration is achieved.

Passage through the tissue is mainly accomplished by the activity of the oral sucker and (where present) an apical cap of spines. The tail is generally cast off before the body completely enters the host. Penetration is accomplished in about 7–8 minutes.

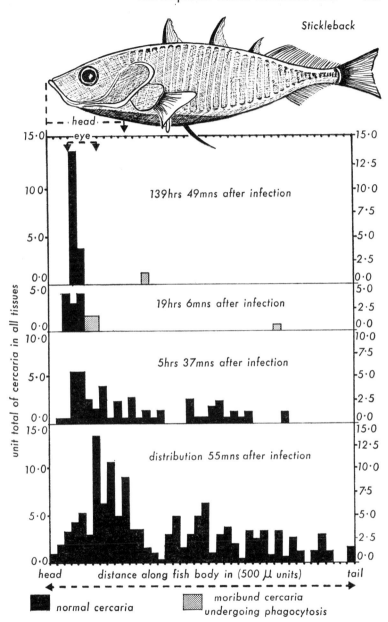

FIG. 45. Distribution of ' Cercaria X ' in the stickleback, *Gasterosteus aculeatus*, during various periods after infection. (modified from Erasmus, 1959)

(*ii*) *Migration*. Once penetration is accomplished, migration to the eye takes place very rapidly. Five and a half hours after infection 27 per cent. of the cercariae are present in the eye, and 72·6 per cent. within 19 hours. Fig. 45 shows the distribution of cercariae in a fish, various periods after infection. After 55 minutes, cercariae have penetrated the entire length of the body, the peak number being in the head, particularly the gill region. After about 5 hours, the overall distribution is essentially unaltered. This general distribution does not persist, however, and after approximately 20 hours the distribution becomes more limited and is concentrated almost entirely within the head region, with a few moribund cercariae in the tail region. After 139 hours the localization is very rigid, the cercariae being confined almost entirely to the lens region of the eye, with a few moribund cercariae undergoing phagocytosis in the central region of the body. In the region in front of the heart, cercariae migrate mainly via the blood vessels, which are readily penetrated here from the gills. In the rest of the body, migration seems to take place via the tissues, and the blood vessels are not especially involved.

There is some experimental evidence that the migration to the eye region is a directional one.[94] Thus, in fish with one eye removed, the cercariae migrate only to the orbit with the eye intact. This is the case even if the cercariae are allowed to enter only the same side of the body as that from which the eye has been removed. If only the lens is removed, cercariae still migrate to the orbit but not in such numbers as when the eye is intact. These experiments suggest that the eye tissue exerts some directional stimuli on the cercariae, possibly of a biochemical nature; but a more detailed examination of this apparent response is clearly indicated.

Into a definitive host

(*i*) *General account.* The schistosomes are the group whose cercariae typically penetrate the definitive host without encystment. Most experimental work has been carried out on *S. mansoni* and *Schistosomatium douthitti*. It has been shown that the penetration gland secretions contain enzymes which enable the attaching cercaria to lyse skin in front of it rapidly. Evidence for the existence of an enzyme or enzymes of the hyaluronidase complex,[307] powerful proteases,[184, 206, 220] a peptidase and a lipase has been presented. Precisely which penetration glands produce which enzymes has not yet been determined. There is no evidence that a true collagenase is secreted, nor is it likely that such would be required by a penetrating cercaria. Thus, although the mammalian dermis consists of a dense matrix of collagen fibres held in an amorphous ground substance in which blood and lymphatic vessels and nerves occur, there is no evidence that a breakdown of collagen occurs during penetration. Entry can easily be effected by penetrating between the collagen bundles.

Release of penetration gland secretions appears to be stimulated by biochemical stimuli from the skin. Mouse skin from which the fat has been extracted does not stimulate this release, but the stimulus is restored on smearing with other extracts of skin. Free fatty acids, such as valeric acid, may be involved.[348]

The point of entry into the skin can be either at the hair follicles or elsewhere, depending on the region penetrated; discontinuities of the surface, such as sulci or follicles, tend to be favoured.[303] In general the path of a penetrating cercaria (termed now a *schistosomula*) seems to follow the path of least resistance with the result that it may even re-emerge at the skin surface! The keratinized layer is the most difficult to penetrate, but once through this, migration is rapid and a schistosomula (see below) is carried to the viscera via the lymphatic or blood vessels, whichever is penetrated first. The normal diameter of the veins and venules does not appear to be a barrier to transport of the schistosomulae since these vessels are usually greatly dilated and congested with blood in the invaded tissues. The rates of penetration and migration have been found to vary considerably with the different types of skin, with the preparative treatment of that skin, and with the number of schistosomulae present.[303] A number of detailed histological descriptions of penetration and migration of cercariae have been given.[121, 124, 297, 303]

(*ii*) *Physiological behaviour after penetration.* Fifteen minutes after penetration, the physiology of a schistosome cercaria, now a schistosomula, undergoes striking changes which are probably related to surface permeability.[304] Thus cercariae are well adapted for life in water, whereas schistosomulae become motionless and vesiculated in water and die rapidly. Conversely, cercariae are not well adapted to saline and serum, becoming˙ shrunken and non-motile after a few minutes in rat serum (table 23), and, although they can survive up to twenty-four hours in certain culture media containing serum and saline (p. 153), they eventually become agglutinated by their tails in this medium.

As shown later (p. 180), cercariae form a remarkable envelope about themselves when immersed in immune serum, a phenomenon known as the *Cercarienhüllen reaction* (CHR). Fifteen minutes after penetration, cercariae lose their ability to form this envelope.

TABLE 23

Comparison of physiological behaviour of cercariae of Schistosoma mansoni *before and 15 minutes after penetration**
(Data from Stirewalt[304])

	Cercariae	Schistosomulae
Water	no effect	toxic
Physiological saline	toxic	no effect
Normal serum	toxic	no effect
Antiserum	formed envelopes (CHR†)	no effect

* Then termed schistosomulae.
† Cercarienhüllen reaction. See p. 180.

These results point to the fact that, after penetration, radical changes occur in the properties of the body layers; the outermost of these either becomes lost or undergoes a molecular rearrangement of some kind. A thin PAS film may sometimes be found around some early schistosomulae,[310] and it is known that a similar film ensheaths cercariae in the snail host; so it is possible that the physiological changes noted are due to modification, rather than loss, of this outer layer. The possible nature of these modifications is not known.

Cercariae eaten directly

An unusual method of infection occurs with some cysticercous cercariae (e.g. of the Azygiidae) in that they are eaten directly by the definitive host. An example of this is *Proterometra dickermani*, the cercariae of which occurs in the snail *Goniobasis livescens*. In this case the cercaria is markedly progenetic, already containing ova (with active miracidia) while still within the snail. The natural definitive host of this trematode is probably a fish, but it is questionable as to how essential a definitive host has become in such a life cycle.

Excystment of Metacercariae

The physiology of excystment has been studied in only a few instances and will be discussed here in detail mainly for *F. hepatica*[76, 78] the metacercaria of which provides useful experimental material. Metacercariae are easily obtained by placing snails in small tubes, each lined with cellophane, and allowing encystment to take place on this surface. It has been shown above (p. 120) that the metacercaria is covered in an inner and an outer cyst each of which is divided into a number of layers. Excystment into the definitive host takes place in two stages. In the first, during mastication or in the gut, the outer cyst wall is removed mechanically; during the second, active excystment from the inner cyst occurs.

The outer cyst may be digested partially by pepsin and further by trypsin in some species, but in *Fasciola* these enzymes alone will not induce excystment even after 23-hour treatment. Excystment from the inner cyst has been shown[76, 78] to depend on the following factors:—

(*i*) a temperature of about 39° C.,
(*ii*) a low oxidation-reduction potential,
(*iii*) a CO_2 system,
(*iv*) presence of bile.

It is especially interesting to note that a somewhat similar system appears to operate in the hatching of certain nematode eggs and in the exsheathing of infective nematode (trichostrongyle) larvae and the hatching of coccidia oocysts. Excystment is thus an *active* process and depends on the stimulation of the larva by specific biochemical and biophysical stimuli occurring in specific sites, and is not merely a passive process involving digestion of the cyst wall by the host secretions. Of the factors mentioned above, the CO_2 system combined with a suitable oxidation-reduction potential, appears to be the most important one involved in the initial—or activation stage—of excystment. Excystment has been reported as taking place, however, in a solution of trypsin–bile, apparently without controlled oxidation-reduction potential and CO_2[352] but in this case too, it is likely that a CO_2 stimulus was inadvertently applied. It can be speculated that the more rapid the excystment of the metacercaria, the greater would be the chance of attachment *in vivo* to the duodenum and subsequent penetration. Although excystment will occur in the absence of reducing conditions, the process is much slower than when these are present (fig. 46). When a suitable oxidation-reduction potential and CO_2 system is provided, hatching can occur within thirty minutes.

It is clear, then, that the larvae possess a hatching mechanism which is ' triggered off ' by a specific set of physico-chemical conditions comparable to those which occur in the duodenum of

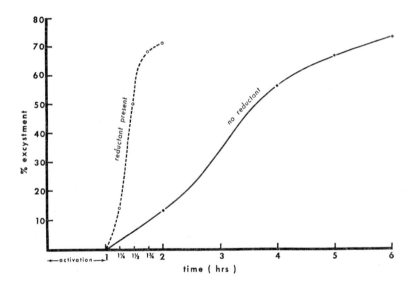

Fig. 46. The metacercariae were activated in Hank's saline gassed with 60 per cent. CO_2/N_2 and, in one series, 0·02M sodium dithionite was added to produce a low redox potential. After 1 hr., the metacercariae were transferred to an emergence medium consisting of Earle's saline containing 10 per cent. sheep bile. The results show that activation is due to high carbon dioxide concentrations and that presence of reducing conditions assists the action of the carbon dioxide and accelerates the process of excystment.

the definitive host. It can be speculated that the hatching process is enzymatic in nature; but there is as yet no unequivocal evidence for this view. Excystment takes place at a particular point in the cyst, the larva escaping through a neat hole in the inner cyst wall (fig. B, plate II), the position of which corresponds to the ventral plug region. Peripheral to this hole is a region which stains in toluidin blue after the juvenile has escaped; this may be indicative of chemical activity in this specific region. It would appear then that the metacercaria is activated by conditions in the host gut which stimulate the parasite to produce an enzymatic secretion which selectively attacks the cyst wall in the ventral plug region to

TABLE 24

Quantitative data on the main acids in some mammalian biles (mg./100 ml.)[293]

Host	Cholic acid	Chenodeoxycholic acid	Deoxycholic acid
Rabbit	?	—	3600
Man	4025	5175	2645
Sheep	3270	?	370
Dog	1555	5	177

form the escape aperture. It will be recalled (p. 70) that the operculum of the egg in *Fasciola* is apparently released by an enzymatic mechanism.

The rôle played by bile in the excystation process is not understood. It is well known that its presence has a synergistic effect on many release processes in helminth life cycles, e.g. the hatching of eggs or the evagination of cestode cysticerci.[293] Bile may play a rôle in determining host specificity for it is recognized that the chemical composition of bile varies markedly from host to host (table 24) and the excystation process of a particular species may require the presence of a specific bile acid at a specific concentration. Again, a specific bile acid may be toxic above a certain concentration. It is interesting, for example, to note the enormous differences between the levels of deoxycholic acid in rabbit and dog bile (3600 mg./100 ml. and 177 mg./100 ml. respectively). It has been found that in certain species of cestodes the cuticle undergoes lysis if particular bile acids are above certain specified levels.[291] The same may hold true for trematodes; and these phenomena could play a part in determining whether or not a metacercaria can become established in a specific host.

Bile has a particularly synergistic effect on the excystation of metacercariae of *Paragonimus westermani*, enabling this process to occur within thirty minutes.[238] It is likely that different species of trematode utilize different types of stimuli to trigger off excystation of their metacercarial stage. In addition to the factors mentioned, any of the normal physico-chemical factors encountered in biological environments could be utilized. A fairly complete list of likely stimuli would then read: temperature, pH, pO_2, pCO_2, bile salts, specific ions, osmotic pressure. Which particular stimulus or stimuli prove to be effective will depend on the nature of the alimentary canal of the host and the position in it normally occupied by the parasite in question.

In some species excystation seems a less complex process than in the case of *Fasciola*. Thus the metacercaria of *Ascocotyle branchialis* excysts in trypsin at pH 8 within 60–90 minutes without further treatment.[325] Comparable results have been obtained with other species. It may be that metacercaria which encyst within an intermediate host, by virtue of being less well protected, excyst more readily than those, such as *Fasciola*, which have cysts with multilayered walls.

Migration Within Definitive Host

General comments

Since the only two means of entry for a trematode into a definitive host is via the skin or a body opening, it is clear that metacercariae (other than those of intestinal species) must undergo a considerable migration through the host tissues before reaching the definitive site of development. Of the 'tissue' parasites, migration to the three main sites, liver, lungs and bloodstream, is mainly dealt with here. The physiological mechanisms underlying the patterns of migration are not understood. Powerful tactic responses may clearly be involved at some stages but the phenomena have been poorly investigated. A metacercaria of an intestinal parasite, for example, presumably responds to a stimulus provided by the host gut conditions, and this enables it to become attached to the intestinal wall. A liver fluke metacercaria, on the other hand, on excystation is similarly stimulated either to penetrate the mucosa (*Fasciola*) or to migrate up the bile ducts (*Dicrocoelium*), depending on the species. Once it has penetrated, a complex pattern of migration follows. A useful technique in these migration studies is the utilization of o·5 per cent. Evans blue, injected intravenously, to detect areas where light bleeding, due to penetration, has occurred. By this means, blue spots along the intestinal serosa can be observed where intestinal penetration has taken place.

Most experimental work has been carried out with laboratory hosts and the possibility exists that the migration pattern may vary substantially in different hosts.

Migration to the liver

The classical studies on this have been carried out with *F. hepatica*[72] with some work on the Chinese liver fluke, *Clonorchis sinensis*, and the lancet fluke, *Dicrocoelium dendriticum*. The route to the liver followed by these species is not the same in each case.

(*a*) *Fasciola hepatica*. This species is unusual amongst trematodes of the liver in that it reaches the liver via the abdominal cavity and not via the bile duct. After excystation, a metacercaria rapidly attacks the mucosa, forms a burrow in it and continues by a tortuous path through the intestinal layers to the abdominal cavity, penetration being completed within 24 hours.[70] During this process, the young fluke is able to break down epithelia, connective tissue and unstriped muscle fibres, and some of these materials can be identified in the caeca of the parasite. Once having penetrated, the flukes apparently browse on whatever tissue is available, occasionally even penetrating lymph nodes. The liver appears to be reached by random wanderings and, once there, parasites attack the liver capsule and form burrows in it. A burrow is only just broader than the body of the trematode and is equal to about six hepatic cells in breadth. As the fluke grows, the liver may be riddled with burrows, so that a substantial loss of host tissue occurs. There is some evidence that young flukes show preference for hepatic cells rather than blood, although some blood is inevitably ingested with this diet.[66, 67, 68] The penetrating flukes eventually reach bile ducts, in which they become permanently established.   Flukes have been recovered from this site in the mouse as early as 24 days after penetrating.

(b) *Clonorchis sinensis.*   A slightly different migration pattern is found in this species; for although penetration of the intestine normally occurs, infection can probably also take place via the bile ducts.   The route in this case appears to be via the portal system[362] and not via the abdominal cavity as in the case of *F. hepatica.*

(c) *Dicrocoelium dendriticum.*   This species does not penetrate the intestine, but infects the liver via the bile duct.[169]

Migration to the lung

This has been studied in detail for *Paragonimus westermani.*[364] Metacercariae begin to penetrate the intestinal wall within 30 to 60 minutes after being ingested and reach the abdominal cavity in 3–6 hours.   Instead of migrating directly to the pleural cavity—as might be expected—they penetrate the superficial layers on the abdominal wall and remain there for about 6–10 days.   During this period, juveniles apparently undergo some growth; and this somewhat unexpected phase in the migration may be obligatory for the purposes of obtaining a particular level of nutrient or a particular growth factor.   After this period, larvae leave the abdominal wall, re-enter the abdominal cavity and migrate through the diaphragm to the pleural cavity and lungs.   The means employed to penetrate the tough muscular wall of the diaphragm are not known;   presumably the young fluke is primarily a tissue feeder, and, as in the case of *F. hepatica*, eats its way through the tissue.   In some hosts, flukes may wander round the abdominal cavity, and grow and reach sexual maturity after penetrating different organs.   In rats, larvae penetrate the abdominal wall but most of the larvae do not migrate back into the cavity but remain in the abdominal wall or migrate to other tissues without becoming sexually mature.   These studies further emphasize the differences in behaviour which may occur in different hosts.

Migration to the eye

A number of trematode species live in the orbital region of birds, under the nictitating membrane.   Migration to this site takes place with amazing rapidity.   Thus, young specimens of

Philophthalmus gralli can be detected in the nasal cavities and beneath the nictitating membranes of chicks 24 hours after being infected orally with the metacercariae,[2] the route apparently being via the crop lumen, oesophagus, nasal cavity and naso-lachrymal duct.   A chemotaxis, comparable with that thought to occur with ' eye ' metacercariae (p. 126) may be involved here.

Migration to the venous system

The broad migration pattern of blood flukes—notably the schistosomes—has been worked out in some detail for *S. mansoni*,[298] although again the physiological basis for such a pattern is not understood.   The actual process of penetration for *S. mansoni* cercariae has already been discussed (p. 127).

The early period of development, after the blood system is reached, appears to be as follows.   In mice, schistosomulae can be found in the lungs after 4 days and reach their highest concentration in this site in 7–9 days (fig. 47).   In the lungs, only a small

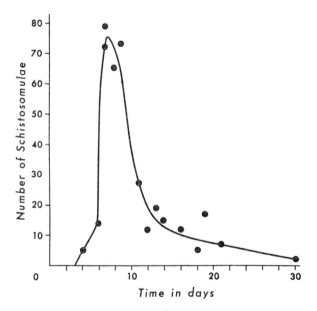

FIG. 47.   Number of stage 1 schistosomulae of *Schistosoma mansoni* recovered from the lungs of mice after various periods after infection with 500 cercariae.   (after Clegg, 1965)

proportion of larvae feed on red blood cells, as is evidenced by the few containing black pigment in the gut. It is remarkable to find that at this stage larvae do not undergo mitosis, so that no growth in size is taking place in the lungs.

After 8 days, schistosomulae can be found in the portal vessels of the liver, a site in which active mitosis takes place. In mixed infections with male and female cercariae, paired worms may be found in the mesenteric and portal veins after about 26 days, but the majority leave the liver on about the 30th day.

Migration to the bladder

A particularly interesting migration is that of *Gorgoderina vitelliloba*, a parasite of the bladder of *Rana temporaria* fairly common in the United Kingdom.

The metacercariae, which may be encysted in alder fly larvae or tadpoles, hatch in the duodenum and pass to the rectum. From there they migrate up the ureter to the kidney where they remain about 30 days. Migration back to the cloaca then takes place and thence to the bladder in which maturation is reached. Some species may remain in the kidney and, in spite of being encapsulated, can reach sexual maturity. If juveniles of *G. vitelliloba* are removed from the kidney and transplanted to the bladder they migrate back to the kidney.[178] This pattern again may be indicative of chemotaxis (to kidney tissue) of one stage of the developing juvenile; but there is no evidence as yet in support of this view.

Growth, Development and Maturation

General comments

Although the growth rates and developmental patterns have been worked out for a number of species, very little is known about the physiological processes controlling growth in trematodes. The rate of growth, as measured by the time required to reach maturity, does not show any clear connection with either the size of the adult trematode or the site in which it occurs. The rate of growth and differentiation, then, are clearly genetically determined in the main, but, as shown below (p. 143), the developmental pattern may differ markedly in different hosts or even in different strains of the same host. There is some tendency for the larger parasites to require longer to mature than smaller ones (in the same site) and, similarly, trematodes in sites with a high nutritional level, e.g. in the upper alimentary canal, tend to grow faster and mature more rapidly than those in sites, such as the caecum, where relatively little food material is available.

Again, species such as the strigeids, which have a ' diplosto-mulum ' stage in a second intermediate host, show a tendency to mature more rapidly than those species whose cercariae either penetrate and develop directly (e.g. the schistosomes) or encyst on vegetation. This may be due to the utilization of materials from the second intermediate host which enable the tissues of the larva to become ' organized ' before being ingested by the definitive host. Many species show ' progenesis ', i.e. gametogeny in the larval condition—a phenomenon discussed further on p. 138.

Growth rate

The most satisfactory method of estimating growth is on a weight basis, as this provides a figure in terms of tissue synthesis; but no data are available on this basis. Some data for growth rates based on length measurements are available.

There appears to be a lag in the growth rate during the first few days of infection but thereafter the rate increases rapidly. In *F. hepatica* in mice, growth is remarkably uniform after the first week of development, the organism approximately doubling its length every 5 days. The growth rates in rabbits and guinea-pigs during the early phases of development are somewhat less.

In *S. mansoni*, between 3 and 4 weeks the doubling time is about 3 days, but between 4 and 5 weeks the rate of growth increases and the doubling time shortens to 2 days.

The growth in size of the various parts of a trematode body is by no means uniform. In *F. hepatica*, the body region containing the main parts of the reproductive system develops to about ten times its larval length, while the suckers only increase about 3–4 times.

Organogeny and maturation

As in other embryos, organogeny and maturation primarily involve the production of a large number of cells by the organism, the differentiation of these into organs and the further maturation of the gonads and associated organs so developed. When discussing maturation it is convenient to divide this process into a series of stages for which recognizable criteria can be applied. Listed below are criteria which have been found to be convenient for strigeid metacercariae,[9] but somewhat modified stages have been adopted for other species.[54, 100]

(a) Stages in maturation

Stage 1: *Cell multiplication.* An outbreak of intense mitotic activity characterizes this first stage of development. This stage cannot be recognized macroscopically but, by means of colchicine (a substance which inhibits cell division in metaphase), mitoses may be easily detected in aceto-orcein squashes.

Stage 2: *Body shaping.* Cell division results in a lengthening of the body, particularly in the posterior region; this results in a bilobed condition typical of the adult strigeid.

Stage 3: *Organogeny.* Not well defined in this species but characterized by the appearance of the anlagen of the genitalia as seen in whole mounts (or aceto-orcein squashes).

Stage 4: *Early gametogeny.* A characteristic and easily recognizable stage marked by the appearance of the early stages of spermatogenesis; the corresponding stages in oogenesis are less easy to recognize.

Stage 5: *Late gametogeny.* Characterized by the appearance of mature spermatozoa in the testes.

Stage 6: *Egg-shell formation and vitellogenesis.* Characterized by the appearance of egg-shell precursors in the cells of the vitellaria. These precursors generally give strong colour reactions (for phenols) with stable diazotates (e.g. Fast Red Salt B) easily recognizable in whole mounts[150] (see p. 60).

Stage 7: *Oviposition.* Characterized by the appearance of fully formed eggs. This stage may generally be readily recognized by compressing living specimens under a cover glass.

Comparable stages for *S. mansoni* are given in fig. 48.

(b) Progenesis in metacercaria

When considering the maturation of a metacercaria in its definitive host, it is essential to take into account the degree of

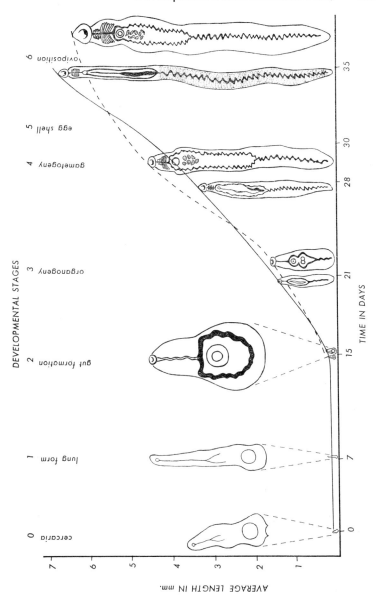

FIG. 48. Growth curve for *Schistosoma mansoni* showing degree of
development after different times. (after Clegg, 1965)

development or differentiation which has already taken place while the organism is still in the metacercarial stage. This varies widely, from metacercariae which contain no genital anlagen (see below) to those which are fully mature and contain eggs.

> The factors controlling ontogeny in trematodes are undoubtedly a complex of interacting internal and external factors and it is difficult to find suitable biological terms to classify the type of development which occurs. Is a ' metacercaria ' with eggs an adult with larval characteristics or a larva with adult characteristics? Development of the former would be ' paedogenesis ' and of the latter ' neoteny ' but the differences between these two phenomena are not always clear. The term ' progenesis ' has also been used to describe advanced development of genitalia in an (apparently) larval form and a metacercaria with genitalia anlagen would be termed ' progenetic '. These terms are somewhat loosely used and definable and applicable only within broad limits. The term ' progenesis ' has been adopted here.

It is possible to draw up a table of metacercariae which represent an almost continuous series of stages of development (table 25). The best known stages are as follows:

(*i*) Metacercariae (or schistosomulae in the case of schistosomes) which show little morphological differences from cercariae and which must pass through all the stages of organogeny before reaching maturity, e.g. *F. hepatica* and *S. mansoni*.

(*ii*) Metacercariae which develop in a second (or, more rarely, in a third) intermediate host and which may grow in size and perhaps alter in shape but which show no organogeny, e.g. *Diplostomum phoxini*.

(*iii*) Metacercariae which have undergone early gametogeny and possess the anlagen of the major genitalia, e.g. *Cryptocotyle lingua*. Such larvae may only require to pass through stages 4 to 7 in order to reach maturity (table 25).

(*iv*) Metacercariae which have undergone gametogeny to a stage where not only are all the major genitalia present but the testes have reached maturity and contain fully formed spermatozoa. Ova are not released from the ovary, however, and eggs are not formed; the vitellaria may show traces of egg-shell precursors, e.g. *Bucephalopsis gracilescens*. Such forms require only to pass through stages 6 and 7 in order to reach maturity (table 25).

(*v*) Metacercariae which become mature in the intermediate hosts. Some species (e.g. *Coitocaecum anaspidis*) produce a small number of eggs while encysted in the crustacean intermediate host, although still retaining a fish definitive host in which large

TABLE 25

Progenesis in metacercariae. A series showing how a progressive degree of maturation occurs in metacercariae of different species

Species	Stages of development*								Prepatent period (days)
	1	2	3	4	5	6	7		
Fasciola hepatica	–	–	D	D	D	D	D		38
Cryptocotyle lingua	I	I	I	?	D	D	D		6
Ascocotyle branchialis	I	I	I	?	?	D	D		3
Bucephalopsis gracilescens	I	I	I	I	I	D	D		?
Microphallus papillorobustus	I	I	I	I	I	I	D		1†
Transversotrema laruei	I	I	I	I	I	I	I	(one egg at a time)	0
Proctoeces subtenuis	I	I	I	I	I	I	I	(many eggs)	0
Proterometra dickermani (cercaria)	I	I	I	I	I	I	I		0

I = In intermediate host;   D = In definitive host.

* See p. 138.   † Eggs appeared after 4 hours in sea water.

numbers of eggs are produced. The metacercaria of *Transversotrema laruei* under the scales of fish produces only one egg at a time and presumably has a definitive host in which more eggs are produced. A further example of a progenetic metacercaria is *Proctoeces subtenuis* which reaches maturity in the organ of Bojanus in the bivalve *Scrobicularia plana* and does not appear to have any other definitive host. In this group could also be included *Proterometra dickermani*, the cercaria of which contains ova while still within the intermediate host (p. 128).

(c) *Factors regulating maturation*

The factors inducing and regulating maturation in trematodes are not understood. As in any embryo, these factors are likely to include those of a nutritional, physico-chemical (i.e. stimulatory) or hormonal nature. Thus, maturation cannot take place until suitable stimuli for cellular differentiation are provided, and a suitable physico-chemical environment and an adequate level of nutrition made available. This question is dealt with further when the *in vitro* cultivation of trematodes is discussed (Chapter 9). A metacercaria may, by accident perhaps, reach a site in the host which is able to support it metabolically but is unable to supply (a) the stimulus to trigger off development (if one be required), (b) a level of nutriment necessary for maturation, (c) suitable environmental conditions under which certain physiological processes, such as feeding or fertilization, can occur. Metacercariae of *F. hepatica*, for example, which penetrate the gut and remain in the coelom, are unable to achieve maturity.

(i) *Nutritional factors.* The importance of the nutritional and hormonal factors is perhaps best illustrated by reference to the life cycle of *S. mansoni*. The maturation curve[54] for this species against time is shown in fig. 48. As was stressed earlier, mitosis in lung forms has not been observed even though several thousands of schistosomulae have been examined. This suggests that the lung lacks some factor or condition for promoting growth. That the important factor may be a nutritional one is suggested by the fact that, once the liver is reached and feeding on red cells commences, cell division begins immediately and the tissues of the worm show large numbers of mitoses.

The importance of the environment is further shown by the fact that when adult schistosomes (*Schistosomatium douthitti*) are experimentally transferred to non-vascular sites, such as the coelom or the anterior chamber of the eye, they regress in a remarkable manner.[120] Thus, females implanted into the eye of rats continue to reproduce normally for about 25 days and to produce eggs. Thereafter, regardless of age at implantation, the ovary and vitellaria become cachectic with loss of morphological integrity and function. Comparable changes occur in the male. After about 50 days, the testes follicles become pale and less discrete, and meiotic activity ceases. After 80–100 days, the testes become reduced in size and may show only isolated groups of non-dividing cells.

Since it is known that—as regards availability of protein and carbohydrates—the eye compares unfavourably with the blood-stream, these results strongly suggest that the de-differentiation of the tissues of the trematodes transplanted into the eye is due more to nutritional than to physico-chemical factors.

It is probable, then, that at least the early stages of maturation depend to some extent on the nutritional level provided by the environment. Thus, the nutritional demands of an egg-producing parasite are generally so great that they cannot be satisfied by the nutriment available in the kind of environment, such as the haemo-coel, usually encountered in the intermediate host. In some cases, however—such as that of *Proctoeces* quoted above—the demands of the adult tissues and the amounts of nutriment available are suffic-iently close to enable maturation to take place.

(*ii*) *Hormonal factors.* The evidence that hormonal factors may play a part in stimulating the sexual maturation of the gonads is shown by the behaviour of *S. mansoni*. The females of this species in the mesenteric portal system of mice will not mature unless accompanied by a male worm. It has been speculated[342] that the male may transfer a hormone either with the sex products or through the integument of the worm, although the spermatozoa themselves may provide the stimulus. Yet another hypothesis put forward is that the females have weak muscles and are unable to feed normally unless paired with the stronger males.[224, 342] This problem raises most interesting questions in cell differentiation.

Influence of host species on morphology of adult

As further experimental work is carried out, it is clear that the morphology of a trematode species may vary with the definitive host species. Thus, the development and morphological variation of *Philophthalmus gralli* in chickens, ducks, rabbits and rats has been studied in detail. It has been found that the arrangement and shape of testes, the location of the acetabulum, the length of prepharynx and oesophagus show significant variations in specimens from different hosts.[48] Comparable results have been obtained with *Diplostomum phoxini*[10] and other species. Clearly, as with free-living organisms, the expression of a gene depends to some extent on the nature of the environment. In the case of a parasite, the different environments provided by different host species may similarly effect its gene expression, with subsequent variation in morphological (and other) characteristics of the parasite. This is clearly an important consideration to be taken into account when questions of speciation are discussed.

9 : *In Vitro* Cultivation of Digenea

General Considerations

It is now widely recognized that many aspects of the physiology and biochemistry of parasitic organisms can be studied satisfactorily only when they are grown in *in vitro* culture free from contamination by other organisms. In other parasitic groups, particularly the bacteria and fungi, advances in knowledge of their physiology have closely followed the development of suitable culture techniques.

A number of terms have been developed to designate particular aspects of culture techniques. Thus, cultivation in the absence of other organisms is termed ' axenic ' (Greek *a*=free from; *xenos*= a stranger). A culture with one other species of organism present is termed ' monoxenic '; when many other species (e.g. of bacteria) are present the term ' polyxenic ' is used. The term ' *in vitro* ' (literally meaning ' in glass ') has long been used to describe a culture involving a liquid or a solid medium in a tube or similar container. An *in vitro* culture is not *necessarily* axenic, although it may be; the widespread use of antibiotics often permits other organisms (e.g. bacteria) to be present in such cultures at a level low enough not to interfere with the development of the parasite.

Basic Problems of Cultivation

Cultivation of trematodes *in vitro* presents a number of special problems due primarily to their complex life cycles. Nevertheless, considerable advances have been made in this field within the last decade; these have been reviewed by Clegg and Smyth.[55] The major difficulties to be overcome may be briefly summarized as follows:

(*a*) Trematodes live in biological habitats, such as the intestine, lungs, liver, blood stream, eye etc., which possess complex physico-chemical characteristics. The nature of many of these habitats is very poorly known. Although the broad characteristics of the host alimentary canal, for example, are known, as shown earlier (p. 22), recent work has questioned previous ideas as to the nature of conditions within the alimentary canal. We have very little accurate information of the precise conditions obtaining in specific regions such as, for example, a crypt of Lieberkühn in which a small trematode might become embedded. The kind of information necessary includes data on the pH, pO_2, pCO_2, oxidation-reduction potential, amino-acid and sugar level, temperature, osmotic pressure and concentration levels of the common physiological ions. Much of this kind of information is exceedingly difficult to obtain with any degree of accuracy, due to the fact that any technique of measurement is itself likely to interfere with the very characteristics (e.g. oxygen tension), it is attempting to measure.

(*b*) Species feed on a variety of biological materials (e.g. blood, bile, mucus, tissue exudates, liver cells, intestinal contents, etc.) whose nutritional properties are complex and difficult to replace by defined media.

(*c*) Many species, such as intestinal flukes, live in non-sterile habitats, so that antibiotic treatment is necessary before axenic culture can be attempted. An alternative solution to this problem is to commence a culture using those larval stages (e.g. metacercariae) which occur in sterile environments.

(*d*) In the natural habitat, the metabolic waste products of a worm are readily removed from the site of their production as a result of the natural circulation of body fluids. A successful *in vitro* method must similarly provide conditions which permit the rapid removal of toxic waste products.

(*e*) The complex nature of trematode life cycles, involving several hosts, means that each stage of development may require different physico-chemical conditions and have different nutritional requirements. ' Trigger ' stimuli (p. 142) may be required before the organism can develop from one stage to another.

Techniques Employed

General principles

In considering so-called ' cultivation ' methods, it is important to distinguish between conditions and media which permit mere *survival* and those which allow growth, development and maturation of trematodes. The term ' survival ' is not easy to define with any precision, and even more difficult to measure *in vitro*. ' Survival ' is here interpreted as involving merely the maintenance of an organism *in vitro*, under conditions which permit the metabolism to operate at a level sufficient to keep its cells and tissues alive but not sufficient to allow growth and maturation in the case of a metacercaria or continued sperm and egg production in the case of adults. By ' growth ' here is meant cell division and tissue synthesis, by ' development ' is meant differentiation of cells into organs, and by ' maturation ' is meant maturation of genitalia to produce normal spermatozoa and eggs. In order to obtain an accurate assessment of these processes *in vitro* it is essential to know a great deal of the pattern of development in the *normal* host, and this should normally be the first step in attempts at *in vitro* culture. Unfortunately, the detailed developmental patterns *in vivo* are known for very few trematodes and are restricted, for the most part, to commonly used experimental organisms such as *Diplostomum phoxini*, *S. mansoni* and *F. hepatica*.

Type of material used

Nearly all *in vitro* work has been carried out using either adult worms, metacercariae or schistosomulae (juvenile schistosomes). The first have the disadvantage that, if they are intestinal forms, they require treatment with antibiotics; and little work has been carried out with adult intestinal organisms. On the other hand, since the upper bile duct is a sterile site, *Fasciola hepatica* may readily be obtained in a sterile condition from an infected liver by opening the host and dissecting the liver with routine aseptic procedures. Since it is difficult to obtain sheep's liver from an abattoir in a sterile condition, this procedure is most readily carried out on experimentally infected animals, such as rabbits.

Schistosomes provide useful material for *in vitro* work since they occur in the blood vessels and liver, from which they can easily be removed aseptically; *S. mansoni* and *Schistosomatium douthitti* both grow well in laboratory mice.

Metacercariae also provide valuable material for experimental work, and many species—particularly strigeids—occur in sterile situations in the intermediate host and have the added advantage that they often occur in great numbers.

Techniques for adult worms

Once sterility is achieved many adult trematodes will ' survive ' for considerable periods in balanced saline solutions, provided suitable conditions of pH, O_2 tension, osmotic pressure and temperature are maintained and a satisfactory method for elimination of waste materials developed.

Studies on ' survival ' techniques are valuable; for they give some indication as to how long metabolic studies could be carried on *in vitro* without abnormal conditions developing. In trematodes these periods appear to be not more than about two hours.

(*a*) *Fasciola hepatica.* Specimens can be ' maintained ' in Hédon-Fleig's saline for 18–34 days at 39° C. in a special culture tube shown in fig. 49. Under such conditions, however, cellular differentiation becomes abnormal within even a few hours, as adjudged by the histological and cytological conditions of the testes.[50] This organ is peculiarly sensitive to abnormal environmental conditions and its cytological examination often provides valuable clues as to the efficiency of the culture methods adopted.

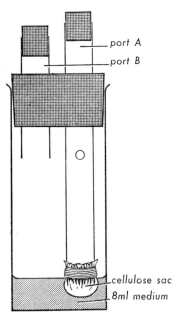

port A
port B
cellulose sac
8ml medium

FIG. 49.   Simple culture tube for cultivation of schistosomulae of *Schistosoma mansoni*. (after Clegg, 1965)

(b) *Haplometra cylindracea.* When aseptically removed from the lung of a frog, specimens of *H. cylindracea* can remain ' alive ' in Hédon-Fleig's saline for periods of 48–105 days.[73] These periods are substantially longer than those given above for *F. hepatica.* As cultivation with this species proceeds, the rate of meiotic divisions drops sharply, which is a clear indication that ' normal 'metabolism is not being maintained.

(c) *Schistosomes.* The maintenance of egg-producing adult worms only will be considered here; techniques for schistosomulae are considered in the next section (p. 152). Adult schistosomes will remain alive for periods of weeks in quite simple media such as serum or serum/Tyrode solution 1: 1 at 36° C., provided the media is changed daily.[229, 260, 261] Metabolizing worms produce quantities of lactic acid (p. 39) and the *p*H drops rapidly. This has a detrimental effect on the worms unless adjusted, since schistosomes are especially sensitive to *p*H changes. The addition of glucose (2 mg./ml.) is beneficial, as is the addition of a small quantity of red and white blood cells or amino acids of the type and proportion found in globulin. That schistosomes digest these cells and require their constituents metabolically is clear from experiments on schistosomulae described below (p. 150). For *S. mansoni*, baboon serum/Tyrode (3: 10) and for *S. japonicum,*

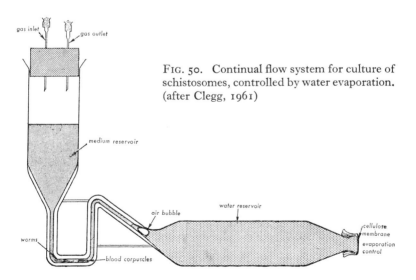

FIG. 50. Continual flow system for culture of schistosomes, controlled by water evaporation. (after Clegg, 1961)

human serum/Tyrode (3: 10) with added glucose and antibiotics, have given the best results.

Both simple tubes or flasks (such as Carrel flasks) and complex apparatus have been used (figs. 50, 51), the latter making provision for easy changing of the medium and observation. Apart from the correct temperature, pH, and osmotic pressure, which are routine requirements, worms seem to be maintained best in a small volume of medium in contact with a large volume of air. Thus, results with quite simple culture vessels have been as satisfactory as those using more complex apparatus.

Although, under the best conditions developed, schistosomes can ' survive ' for periods of many weeks or even several months,[260] it is doubtful if a metabolism approaching that *in vivo* can continue

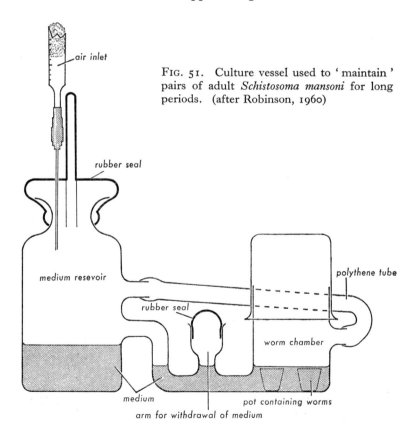

Fig. 51.   Culture vessel used to ' maintain ' pairs of adult *Schistosoma mansoni* for long periods.   (after Robinson, 1960)

for more than a few days.   In early experiments, worms
failed to produce viable eggs but, with the media and methods
described above, worms remained *in copula* and produced viable
eggs which hatched out active miracidia.[228]   Although these
miracidia penetrated the snails *Australorbis glabratus* and *Oncomelania hupensis*, larval stages failed to develop in the molluscs—a
result showing that they were abnormal in some way.   They may
possibly have lacked some growth factor without which further
development cannot take place.   Some attempts to develop a
completely synthetic medium, which would allow more precise
analysis of growth requirements, have been made.   A medium
containing 26 amino acids, a series of enzymes, coenzymes, nucleic
acid derivatives, vitamins, glucose and gluconates in a balanced
saline solution has been developed.[280]   This medium has proved
to be adequate only for short term maintenance and in it females
can produce eggs for a period of twenty days.   Eggs obtained in
the experiments were almost 100 per cent. non-viable although an
occasional one showed flame-cell activity, indicating that some
embryonic development had occurred.   Miracidia were never
obtained, so that results in such media were not as satisfactory as
those obtained in the Tyrode/serum/glucose mixture.   These
experiments, being necessarily empirical, have given us some very
useful techniques for the maintenance of schistosomes *in vitro*
but so far they have provided little detailed information about the
nutritional requirements of adult worms or the precise effects of
various physico-chemical factors (e.g. O_2 tension) on their
metabolism.

Techniques for juvenile worms

General comments

The advantages of using juvenile worms for culture work are
several.   Firstly, as metacercariae or schistosomulae they can
readily be obtained in an aseptic condition from the intermediate
host.   Secondly, if development is proceeding satisfactorily, they
pass through easily recognizable stages, so that growth responses
to change of media or conditions can be more readily assessed than
if adult worms are used.

Schistosomes

A number of attempts to culture *S. mansoni* to maturity have been made using schistosomulae at various stages of differentiation.[35, 51, 54] The development of *S. mansoni*, which requires 35 days to mature in a mouse, can conveniently be divided into a number of stages which differ to a small extent from those listed on p. 138. Briefly, the pattern of development consists of cell multiplication and gut formation, followed by the appearance and maturation of the genitalia; the whole process is, of course, accompanied by a marked growth in size (fig. 48).

Cultivation of 7-day old schistosomulae. Schistosomulae of *S. mansoni* recovered from the lungs 7 days after infection (stage I) form especially valuable starting material for *in vitro* culture work, because somatic growth has not started, i.e. there are no mitoses (p. 138) and the worms are all at the same stage of development.

Culture vessels and conditions. Good results can be obtained using a standard 20 ml. McCartney-type bottle with a flat internal base and a rubber stopper. A modified culture tube is shown in fig. 49; it consists of a tube whose lower open end is covered by a small sac of cellulose membrane (dialysis tubing), and so has the advantage of permitting waste products to diffuse readily from the vicinity of the worms. Although this diffusion would be greater if the outside medium was agitated, schistosomes (unlike strigeids, p. 156) do not develop well in agitated cultures (p. 154). In both these culture vessels the 1 per cent. red cells, used in the medium described below, sediment to form a thin layer at the bottom of the tube. It is in this thin layer that the schistosomulae develop. Since the *p*H of portal blood is constant at 7·4 it is not surprising to find that *p*H is especially important and that development is inhibited by even small variations from this figure.

Medium. The most satisfactory medium so far developed is given in table 26. It consists essentially of inactivated rabbit serum, Earle's saline, lactalbumin hydrolysate, glucose and red cells. The addition of other nutrients, such as chick embryo extract or liver extract, which have proved useful in tissue culture or culture of other parasites, particularly nematodes, has a detrimental effect on development and these are, therefore, not suitable additions to the medium.

Results. In Clegg's No. II medium under the 'best' conditions quite a remarkable degree of growth can be obtained and schistosomulae will consistently develop at least to stage 4 (fig. 48), characterized by the appearance of sperm in the male, and occasionally to stage 5, characterized by the appearance of egg-shell protein in the vitelline cells.[54] The latter do not develop further, however, and, although pairing will occur regularly in this medium, fully formed eggs are not produced. Cultured schistosomulae reach stage 5 within six weeks but worms do not grow as large or as rapidly as those grown in a mouse. The reduction in the rate of growth is most marked in the later stages of culture, and growth stops altogether after six weeks.

TABLE 26

Culture medium for growing schistosomulae of S. mansoni *to near-mature adults (Clegg[54])*

Clegg's Medium No. II

Inactivated rabbit serum ⎱ Equal parts
Earle's saline ⎰

Plus: 0·5% lactalbumin hydrolysate
0·1% glucose
1% rabbit red cells
100 units/ml. penicillin
100 μgm./ml. streptomycin

The development of a schistosome from a 7-day-old schistosomula to a near-adult represents a considerable advance in culture technique and should enable the physiology of growth during this period to be accurately studied.

Cultivation of cercariae in vitro. Experimental attempts[54] to develop adult schistosomes commencing with cercariae released from snails (instead of schistosomulae, as above) have revealed that *cercariae must penetrate the skin before they can begin to develop into schistosomes.* This result is not surprising, for it has already been shown (table 23) that marked physiological changes occur in a cercaria after penetration. It may be that the release of the penetration glands (p. 96) is the chief 'trigger' involved. If cercariae freshly released from snails are cultured in Medium II, they swim freely for the first 10 hours in much the same manner as they would in water. After 24 hours however, a type of agglutination occurs and cercariae are held by their tails in groups of 7–15. After 48 hours in this media cercariae die.

If released cercariae are allowed to penetrate some mouse skin and then collected immediately afterwards, they may be cultured in Medium II in the same manner as 7-day lung schistosomulae. After 5 days in this medium, they begin to feed on red blood cells. Subsequent development of schistosomulae from cercarial stages follows almost exactly the pattern described for schistosomulae, except that development does not extend beyond stage 4.

Attempts to improve cultivation conditions

The failure of culture techniques described above to produce fully matured worms could be due to a number of causes. The major ones could be lack of a nutritional factor or factors, or unsuitable physical conditions such as local accumulation of excreted metabolites, pCO_2, pO_2 etc. As mentioned, the addition of the various growth stimulants such as chick embryo extract or liver extract has generally proved inhibitory, and not stimulatory, to further development. Agitation of the cultures during incubation by various techniques proved generally not to be beneficial to development, although one single worm in an agitated culture reached full development and produced an egg. It is thought that this worm may have been subjected to some unusual condition during agitation—such as being held in a fold of the cellulose sac —and that 'freak' perfect conditions for development resulted. This one result suggests that failure to develop to maturity may be due to lack of precisely suitable physical cultivation conditions during the final stages of development rather than to failure to satisfy the nutritional requirements. It is possible, of course, that unsuitable cultivation conditions could indirectly affect the availability of nutriment by failing to provide the conditions under which normal feeding can take place.

Continuous flow apparatus. An ingenious method for cultivating schistosomes and possibly other organisms in a continually flowing medium has recently been developed[52] and is shown in fig. 50. The beauty of this simple apparatus is the ease with which the rate of flow can be controlled. Evaporation of water vapour at the surface of the cellophane membrane causes the culture liquid to flow along the narrow bore tube. The rate of flow is controlled

simply by painting the cellophane membrane surface with a cellulose paint (nail varnish!); the rate of flow is then proportional to
the area left uncovered. A fan in the incubation chamber may also
be used to alter the rate of evaporation and hence the rate of flow.

In vitro *cultivation of* Paragonimus westermani

Some preliminary experiments with the metacercaria of the
lung parasite, *P. westermani*, which is a slow growing parasite,
suggest that this tissue parasite may thrive better on embryo
extract than schistosomes. Thus, in balanced saline plus chick
embryo extract and red blood cells metacercariae have grown from
o·6 mm. × o·26 mm. to 2·0 mm.—2·5 mm. × 1·5 mm. in 98 days
and the primordia of the genitalia developed. 363a.

Cultivation of strigeid trematodes

Metacercariae of strigeid trematodes offer exceptionally good
material for *in vitro* work in that (*a*) they occur commonly in fish
or amphibia and may sometimes be obtained in large numbers,
(*b*) some species are found free in sterile environments so that
neither excystment nor sterilizing procedures are necessary before
experimental work can be commenced, and (*c*) some species mature
rapidly.

Progenetic metacercariae. Some species exhibit marked
progenesis with partial development of the genitalia in the metacercarial stage. Thus, in the metacercaria of *Posthodiplostomum*
the genitalia are differentiated up to the early stage of gametogeny
(stage 4; fig. 48). On theoretical grounds, therefore, differentiation to the final maturation stages should be rapid and, if endogenous reserves are sufficiently large, maturation could take place
in vitro, even in saline solution, assuming that the correct physico-
chemical conditions of *p*H, temperature, osmotic pressure, gas
phase etc. simulating those in the host, were provided. In the
species examined, these food reserves are not, in fact, sufficient
to permit maturation to occur. Thus, when the metacercariae
of *Posthodiplostomum cuticola* are removed from fish and incubated
in physiological saline, differentiation up to late gametogeny, but
no further, occurs.[30]

Undifferentiated metacercariae. An undifferentiated metacercaria is one in which the anlagen of the genitalia are not developed.

Development of such a form to an adult strigeid thus involves considerable tissue synthesis; the organism must move from stage 0 to stage 7 (fig. 48).

In Europe, one of the commonest strigeids with metacercariae of this type is *Diplostomum phoxini*, whose metacercariae occur in the brain of the freshwater minnow *Phoxinus phoxinus* and whose adult stages occur in water birds. In the laboratory, maturation may be conveniently carried out in ducks. It is a comparatively simple matter to remove metacercariae from the brain of a minnow with aseptic precautions and transfer them to a suitable medium in a suitable container.[9] Any one of a number of conventional vessels has been found to be satisfactory: 25-ml. screwtop vials are convenient. Cultivation is carried out at 40° C. and the culture vessels are shaken briefly every 1–2 minutes by an electric shaker. This simulates intestinal movement and, in contrast to schistosomes, provides more suitable cultivation conditions; the movement also assists in the diffusion of waste materials from the vicinity of the worms.

In a medium of yolk (10 ml.), albumen (2 ml.) and balanced saline solution (2 ml.), the metacercariae may be brought to near-maturity in that active sperm but abnormal eggs are produced; normal eggs have not yet been obtained by this technique. This rather complex medium is of a viscous nature and when metacercariae are cultured in it, their intestines rapidly fill. The nuclei of the juvenile tissue begin to divide rapidly and the genitalia appear. The testes produce masses of active sperm and, by 84 hours' culture, eggs finally appear.[9] Eggs produced by this technique are thin-shelled and morphologically abnormal. Correlated with the failure to develop normal egg-shells is the poor development of the vitellaria in which the egg-shell material is synthesized (p. 60). When yolk is omitted from this culture medium—leaving only albumen and gluco-saline—very little development occurs in the posterior, genitalia-containing region of the worm.

It is difficult to say why this complex mixture is comparatively successful; there are several possible explanations, all or some of which may be true. (*a*) It may provide bulk or trace nutrient materials; (*b*) it may provide a suitable, well-buffered, physicochemical environment; (*c*) it may have a consistency which makes it especially suitable for being taken into the gut. Attempts have been made to replace the yolk constituent in the medium by synthetic media.[363] A mixture of serum+amino acids, when added to the albumen and gluco-saline, has been found to act as a fairly adequate substitute and allows the same level of maturation to take place. The addition of yeast extract to the above medium favours sperm production, and it is clear that the nutritional requirements of sperm production are contained in aqueous yeast extract, which *inter alia* contains B vitamins and a high concentration of amino acids. The serum appears to contribute one or more essential metabolites not present in yeast extracts.

Two active fractions have been found in the yeast component: (1) a dialyzable fraction, the effect of which can be replaced by amino acids, and (2) a non-dialyzable fraction which is resistant to hot acid and alkaline hydrolysis, and is basically charged.[358] Part of this non-dialyzable fraction is probably vitamin B_6 (pyridoxine) and its effectiveness may be related to the part played by pyridoxal phosphate in the activation of a number of enzymes such as transaminases, decarboxylases, alanine racemase, dehydrases and desulphydrases, where it acts as a co-enzyme. These are important enzyme systems, and it is easy to see the important rôle this substance could play in the maturation of the fluke.

The position regarding the metacercariae of *Diplostomum phoxini*, then, is that relatively satisfactory techniques are available for cultivating it to near-maturity, but that very few of the constituents of the medium are defined, and normal eggs, capable of embryonation and hatching, have yet to be obtained. These results suggest that, as with schistosomes, the final stages of maturation make nutritional demands which present media and techniques do not satisfy. The nature of the deficiencies in the media are, at present, unknown.

Cultivation in chorioallantois of chick

Excysted metacercariae of *Philophthalmus* sp. have been matured in the chorioallantois of developing chick embryos.[100] Such metacercariae reach maturity in 14–20 days, as compared with 13 days in the bird host. While this is not strictly *in vitro* culture, it does provide a useful technique for experimental study of growth which may be applicable to other species.

10 : Physiology of the Host-Parasite Relationship

I. Tissue Reactions

General Considerations

Problems of the host-parasite relationship

In the case of a free-living organism, a knowledge of the nature of the biotope and the changes to which it is subjected is fundamental to an understanding of the physiology of the organism. Similarly, with parasitic animals a knowledge of the biotope provided by the host and the interactions between host and parasite is a fundamental prerequisite for our understanding of the physiology of parasites.

It has been pointed out elsewhere (p. 146) that, in general, the nutritional and physico-chemical properties of the various vertebrate biotopes utilized by trematodes are poorly known. Likewise, little precise data are known regarding the actual physiological processes being carried on by the host tissues within these potential parasite sites. Thus even such a fundamental mechanism as that by which carbohydrate is absorbed from the duodenum remains largely a mystery (p. 22).

It is not surprising, then, to find that knowledge of the physiology of the host-parasite relationship is not extensive; for this is a field in which much remains to be done. In dealing with this subject we need to consider (*a*) how the host reacts to the presence of a trematode parasite, (*b*) what factors are involved in determining host specificity for particular species and (c) what influence, if any, variation in the host physiology has on a parasite. Many of these questions raise complex issues, which are common to the pheno-menon of parasitism as a whole and beyond the scope of this book. Immunological reactions peculiar to the trematodes, however, are of particular interest and will be treated in some detail.

General host reactions

Much of the basic work in this field has been carried out with vertebrates, but there is some information on tissue reactions in molluscs and insects. When a host is invaded by a foreign body which, in some way not understood, it can distinguish as being 'non-self' material (see p. 170), it responds by bringing into action defence mechanisms which can be broadly considered under two categories, which overlap to some extent:

(*a*) tissue reactions,
(*b*) immunity or resistance.

Detailed consideration of these mechanisms in vertebrates as they apply to helminth parasites has been covered in several texts or reviews[290, 294, 295, 296] and will be dealt with only in outline here. In general, it can be said that a tissue reaction tends to be localized in the *immediate* site of the host-parasite interface; it usually appears rapidly after initial contact and often disappears after the invading organism has left or has been destroyed. Immunity (=resistance; see below), on the other hand, is a more generalized effect, originating in organs or systems far removed from the site of initial host-parasite contact, and may be considered to be a reaction of the whole body to invasion by foreign material. Immunological reactions usually develop more slowly than tissue reactions but generally persist for a longer period, sometimes (in the case of certain viruses or bacteria) apparently throughout the life of the host. Theoretically, 'immunity' implies *freedom* from invasion by an organism, whereas 'resistance' implies that the defensive mechanisms of the host may not be completely successful in repelling the invasion of 'foreign' material. However, the terms are widely used as being synonymous and are generally so used in this book.

Vertebrate Host Tissue Reactions

Inflammatory reaction

The effect on the tissues of the presence of a parasite is to invoke an *inflammatory reaction* which is an important part of an organism's defence mechanism against foreign material within the body. An excellent general account of this process is given by Sprent.[296] If the amount of foreign material is small (e.g. an egg), it will gradually be surrounded by phagocytic cells and immobilized by the deposition of collagenous tissue around it. If a large amount of foreign material is present, the reaction is much more severe; this is partly a result of the mechanical irritation and also of the effect of released metabolic products on the host cells.

The first effect is a local dilation of the capillaries (vasodilation) which is brought about by a local nervous reaction and the release from the affected cells of substances such as histamine. This results in an increase of blood supply to the affected area accompanied by an increased permeability of the capillary walls and the passage of protein materials from the blood into the tissue fluids. In the region of the invaded tissue, the vessel walls appear to become sticky and leucocytes adhere to them. The leucocytes then proceed to infiltrate through the vessel walls and collect in large numbers at the site of invasion.

The leucocytes appear to be attracted to the injured area by specific substances (e.g. leucotoxine), which are probably released from damaged cells. Heterophil leucocytes, which are actively phagocytic, are the first cells to migrate to the infected area. They are followed by lymphocytes. Their arrival may be associated with a change in the local pH, brought about by an accumulation of lactic acid, probably of both host and parasite origin. This change in pH is apparently more favourable to lymphocytes than to heterophils. It is thought that lymphocytes are relatively undifferentiated cells which can become transformed into mononuclear cells or fibroblasts; the latter form the fibrous outer 'capsule' (see p. 123) so characteristic of many metacercarial cysts in intermediate hosts.

Reactions to metacercariae and mesocercariae

These range from no reaction at all to one which results in the metacercaria being completely encased in a fibrous capsule. In general, the sites where no marked reaction occurs are those in which there is a dearth of connective tissue cells which could be utilized for encasing a parasite. Thus, a number of metacercariae (e.g. *Diplostomum spathaceum*) occur in the humor or lens capsule of the eyes of fish, a site in which no host reaction is provoked. Other sites where no reaction occurs are the lymph spaces between the tissue of tadpoles, frogs and frog-eating snakes where the mesocercariae of *Alaria mustelae* and *A. marinae* live. These larvae, not being provided with cystogenous glands and occurring in sites lacking suitable connective tissue elements, are not encased in a ' cyst ' of either parasite or host origin.

The brain, too, is a site well known to be somewhat immunologically inactive; and some fish species can be invaded by enormous numbers of metacercariae without showing much tissue reaction. Thus, the brain of the minnow *Phoxinus phoxinus* can house up to 1200 metacercariae of *Diplostomum phoxini* without affecting the behaviour of the fish to any noticeable extent.[256] In this case, however, some cell proliferation does occur in the vicinity of the parasite and loose coats of connective tissue are formed around them.[4]

In sites with abundant connective tissue cells, such as the muscles of fish, the host tissue reacts strongly against a larval trematode and forms a fibrous capsule round it. The process whereby it does this does not appear to differ significantly from that of a normal inflammatory reaction to foreign tissue; it has been well described in fish, against the cysts of *Uvulifer ambloplitis*.[142] In most cases, within 2–4 days the trematode secretes a hyaline cyst wall, the structure of which may be more complex than it appears. This is rapidly surrounded by host cells, especially at the poles, so that the developing capsule appears lemon shaped. Further cells accumulate and fibres of loose connective tissue appear. As development continues, fibroblastic tissues come to form an outer coat. In many species, melanophore-like cells appear in the loose connective tissue of the cyst and become incorporated into the cyst wall giving the whole a pigmented appearance (plate III, fig. A). The pigment-bearing cells are probably true melanophores, since injection of adrenaline causes some contraction of pigment, as it does in normal melanophores in fish skin. Some melanin may also possibly be formed by the release of tyrosinase from the parasite and its subsequent action on the adjacent tissue (see p. 62). It is worth noting that there are a few species of cyst-dwelling adult trematodes also. Thus, the unisexual adults of *Köllikeria filicollis* occur in ' cysts ' in the branchial cavity of fish.[357] The parasites secrete a cyst which is surrounded by host tissue cells.

Reactions to developing schistosomes

Schistosomes constitute a special case, since development after penetration of the cercaria is direct, so that the definitive host is subjected to all the developmental stages from cercaria to adult.

(*a*) *Cercariae.* After penetration, the *stratum corneum* in the skin undergoes some thickening so that more resistance is offered to cercariae attacking later.[302] Even in ' normal ' hosts, some local resistance of a cellular nature to cercarial penetration may develop (fig. 56). A cercaria of *S. mansoni*, for example, normally penetrates and makes its way to the liver without invoking an evident cellular response; if, however, for some reason its passage is slowed down, leucocytes accumulate and the parasite becomes surrounded by host cells.

In ' abnormal ' hosts, cercariae may penetrate and undergo partial or complete development (see below) or they may be stopped at the skin or subcutareous level. Thus it is well-known that the cercariae of many species, specially bird schistosomes, penetrate the skin of humans and cause ' cercarial dermatitis ' or ' swimmer's itch '.[202, 203] The tissue response may be slight or severe. In unexposed subjects, a prickling itch develops within 10–15 minutes, and 1 mm. macules (inflammatory patches due to local vasodilation) develop. In previously exposed subjects, the skin becomes ' sensitized ' (p. 174) and the reaction is more severe. The penetrating cercariae appear to be immobilized, to die slowly and to undergo lysis. As some precipitates appear around the cercariae, these phenomena may be the result of an antibody reaction (see p. 171); only in the case of ' sensitized ' subjects are the parasites invaded by phagocytes.

(*b*) *Schistosomulae.* Schistosomes present an interesting example of the range of developmental patterns which can occur in so-called 'abnormal' hosts (see p. 144), and of the part played by tissue reactions in limiting this development. Thus, in one experiment, the development of *S. mansoni* was studied in 13

TABLE 27

Relative susceptibilities of mammals to infection by Schistosoma mansoni

N=normal; S=stunted; VS=very stunted; G=good; P=poor;
— to +++=none to many

(Data from von Lichtenberg, Sadun and Bruce[347])

Suscepti-bility (order)	Host species	Size	Adult worms			Eggs			Viabi-lity	Infect-ivity
			Location			Location				
			Liver	Portal veins	Mesen-teric veins	Liver	Intes.	Stools		
I	Woodchuck	N	—	+	+++	+	+	++	G	G
	Lab. Mouse	N	—	+	+++	+	+	++	G	G
	Squirrel	N	—	+	+++	+	+	++	G	G
II	Opossum	S	++	++	++	+	+	+	P	P
	Meadow Vole	N	++	++	+	+	+	+	P	P
III	Nutria	S	++	++	++	+	+	±	P	—
	Chipmunk	S	++	++	++	+	+	—	P	—
	Skunk	N	++	++	+	—	—	—	—	—
	Racoon	S	++	++	+	+	—	—	—	—
	Cottontail Rabbit	VS	+++	—	—	—	—	—	—	—
IV	House rat	VS	+++	—	—	—	—	—	—	—
	Musk-rat	—	—	—	—	—	—	—	—	—
	Fox	—	—	—	—	—	—	—	—	—

species of mammalian hosts. The general order of host suscepti-bility is given in table 27, from which it is clear that a wide range of parasitological patterns occur. Hosts may be broadly divided into four groups—'very susceptible', 'partly susceptible', 'poorly susceptible' and 'resistant'; but these categories are not homo-geneous. In the 'very susceptible' group adults migrated to the mesenteric veins and large numbers of eggs were produced. In this group the woodchuck was found to be more susceptible than the laboratory mouse, although the latter is much used for experi-mental purposes.

In the 'partly susceptible' group there was a tendency towards stunted development, and, although in some cases egg production was relatively high, many eggs underwent degeneration. The 'poorly susceptible' group consisted of the meadow vole, chipmunk, racoon and nutria. In these hosts there was marked stunting of adults and almost all eggs degenerated; granulomas were predominantly small.

In the 'resistant' group the parasites were not successful in completing their life cycles but marked differences were observed between the various species. The least resistant was the rat, in which a few stunted adults developed but no eggs were

TABLE 28

Degree of inflammation elicited by migration and development of schistosomulae in susceptible and resistant animals

n=no observation; − to +++ = none to many

(Data from von Lichtenberg, Sadun and Bruce[347])

Category	Host Species	Skin—hours after exposure							Type of inflammation
		1	5	8	24	48	72	120	
Very susceptible	Mouse	±	n	++	++	n	±	n	Weakest; nearly normal at 72 hrs.
Very susceptible	Squirrel	+	++	+++	+++	++	+	+	Strong early reaction. Spotty haemorrhage.
Resistant	Musk-rat	±	++	n	+++	+++	+++	+++	Late stage focus in dermis.
Totally resistant	Fox	+	n	+++	+++	+++	+++	+++	Strongest, late stage diffuse.

formed. In the musk-rat, numbers of schistosomulae reached the lungs where some evoked a tissue response and were trapped. In the fox, resistance was almost complete and the schistosomulae were destroyed in the dermis and the subcutaneous tissue. The tissue response in some of these hosts is shown in table 28, and it is clear that, as expected, the tissue reaction is greatest in the case of the more resistant hosts and least in the more susceptible hosts.

Reactions to adult trematodes

Only the tissue reactions which occur in main sites occupied by adult trematodes will be dealt with here; the gross pathology of adult trematode infections is dealt with in major works on medical and veterinary parasitology. As with larval forms, the degree of host response depends on the extent to which the parasites make contact with the somatic tissue. The major factor concerned is probably how far the host has an opportunity of recognizing the parasite as ' foreign ' and bringing its defence measures into play at a tissue or serological level. Thus tissue forms will, in general, provoke a much greater reaction than intestinal forms, unless the latter penetrate the mucosa or make a particularly intimate contact with it.

(*a*) *Liver.* In general, liver tissue reacts actively to the presence of trematodes such as *Fasciola hepatica*, *Clonorchis sinensis* and *Dicrocoelium dendriticum.* These cause a progressive hyperplasia (=excessive multiplication of cells) of the bile ducts with adenomatous (=tumour-like) proliferation and heavy infiltration of connective tissue. In severe infections there may be an extensive replacement of liver parenchyma by connective tissue, which may result in portal cirrhosis. Eggs of schistosomes reaching the liver, particularly in *S. mansoni* and *S. japonicum* infections, also provoke fibroblastic proliferation.

(*b*) *Lung.* The presence of *Paragonimus westermani* provokes a neutrophilic and eosinophilic reaction in the tissues with the formation of a thick fibrous capsule.

(*c*) *Urinogenital system.* The urinary and genital systems, too, react actively to the presence of foreign tissue. The most studied infection of these systems is with the eggs of schistosomes, especially *S. haematobium.* Invasion leads to inflammatory reactions with progressive changes in the bladder and genitalia. The cellular irritation invoked by the presence of schistosome eggs undoubtedly results in a predisposition towards malignancy. The disease of schistosomiasis has long been associated with carcinoma of the urinogenital system and may be proved to be one cause of this condition, but the relationship between the two has not been unequivocally established.

(*d*) *Alimentary canal.* In general, intestinal lesions produced by trematodes are relatively negligible, but some forms do produce localized foci of inflammation at the site of attachment. In this respect it is recalled (p. 22) that the mucosal cells have a very short ' turnover time ', so that the period for cells to show any reaction is limited.

For example, the large intestinal fluke of man, *Fasciolopsis buski,* and the stomach fluke of cattle, *Paramphistomum microbothrium,* can cause irritation and inflammation to the intestinal mucosa if present in large numbers.[75] In the case of the latter species, it is only the immature forms which are concerned. These occur in the duodenum but migrate into the rumen and reticulum, where, as adult worms, they apparently have little effect on the host tissues.

(e) *Miscellaneous tissues.* The foregoing are the main sites in which trematodes commonly occur; but some species occur in aberrant sites (such as *Spirorchis haematobium* in the heart of a turtle) and others show a tendency to wander. It is probably true to say that in most sites an inflammatory reaction of some type is invoked. The extent of this will depend on many factors not fully understood, but the ' strain ' of the host as well as the ' strain ' of the parasite often plays an important rôle (see p. 5).

Invertebrate Host Tissue Reactions

The initial defence mechanisms of invertebrates form a comparatively neglected field of study, but one which is of particular importance to the understanding of trematode ecology, distribution and possible control. Several useful reviews are available.[12, 41, 136, 274, 300, 326]

It is of interest to note that the phenomenon of phagocytosis was first observed in 1884 by Metschnikoff in an invertebrate, the freshwater cladoceran *Daphnia,* and has since been seen in numerous other species of invertebrates. Some evidence is also available to demonstrate the presence of humoral antibodies against microorganisms in the blood and body fluids of invertebrates.

Although metacercariae occur in a number of invertebrates other than molluscs, their early larval development occurs exclusively in molluscs, and the tissue reactions of molluscs in particular are examined here. Reactions of insects to metacercariae are dealt with later (p. 169).

Molluscs

Host tissue reactions in molluscs range from no detectable reaction to an intense cellular response. They can be discussed under three main headings: (a) phagocytosis, (b) encapsulation and (c) leucocytosis. There is some evidence of humoral immunity in molluscs and this is dealt with in Chapter 11.

(a) *Phagocytosis.* This is carried out in molluscs by amoeboid blood cells which have phagocytic properties. These have been termed *phagocytes, amoebocytes, haemocytes* or *leucocytes* by various workers. The functional term haemocyte is used here. It has been clearly shown that indian ink particles injected intracardially into the oyster *Crassostrea virginica* formed emboli (clots) which subsequently became broken up and the particles readily phagocytized. A wide range of biological materials including the following have been found to be attacked and ingested by the phagocytes: starch granules, bacteria, spores, bovine haemoglobin, human serum albumin, diphtheria toxoid, normal erythrocytes of fish, birds and mammals and even duckling erythrocytes enclosing the relatively indigestible malaria pigment. Material, such as erythrocytes, which is digestible by cellular activity, can be digested by molluscan haemocytes. Indigestible material, such as pigment, on the other hand, is eliminated from the body of oysters by the passage of the haemocytes bearing them through the epithelial lining of the alimentary canal or through other organs to the external aquatic environment. Phagocytosis apparently takes place very rapidly; yeast cells, for example, when injected into the snail *Australorbis glabratus* are phagocytized within ten minutes and the process of migration of haemocytes to the outside can go on for several weeks.

It is interesting to note that fresh *A. glabratus* tissue when implanted into *A. glabratus* provoked no fibroblast response (see below) and only a transient haemocytic infiltration, the latter probably in response to damaged cells. Formalin-fixed snail tissue, in contrast, evoked fibroblast response and large particles, such as artificial latex spheres, which cannot be phagocytized, were encapsulated.[326]

(b) *Encapsulation.* Encapsulation serves as an efficient defence mechanism for material which cannot be disposed of by phago-cytosis. It involves the enclosure of the foreign material in a fibrous capsule formed from fibroblasts in a manner somewhat similar to that in which they form in vertebrate tissues. The phenomenon may be elegantly illustrated by reference to infections of *Schistosoma mansoni* in two strains of snails. Thus, when the Puerto Rican and Brazilian strains of *Australorbis glabratus* are exposed to miracidia of this schistosome, the latter do not become infected, but the former become heavily infected. Within the Brazilian strain there is a rapid tissue reaction, and a fibrotic type of encapsulation occurs, followed by degeneration of the parasites. Young snails of the Brazilian strain, however, are susceptible, so that an age-resistance factor clearly operates.[230, 231, 232] A comparable situation occurs with larvae of some pseudophyllidian cestodes. In this case, copepodid larvae are often more susceptible to infection by coracidia than adult copepods.

Several other experiments have been carried out[41] with species of trematodes whose miracidia have been allowed to infect an ' unnatural ' molluscan host. In most cases there was a marked host tissue reaction within twenty-four hours, followed by encapsulation. Encapsulation may not result in the immediate death of the encapsulated larval tissue but, in general, this rapidly degenerates.

Encapsulation occurs in some species apparently without harming the parasite, but, on the contrary, supporting and protecting it. Thus, in *Glypthelmins quieta* the egg hatches in the gut and the miracidium penetrates the gut epithelium, and the mother sporocysts develop between it and the underlying basement membrane. The tissue response invoked results in the multiplication of the basement membrane cells to form cellular membranes, termed *paletots*, which invade the mother sporocyst and form round individual daughter sporocysts. The daughter sporocysts thus remain in a compact mass held together by their paletots.

(c) *Leucocytosis*. This type of host response is characterized by a marked increase in the number of white blood cells. The phenomenon has been but little studied. It has been reported[41] that the number of leucocytes in oysters was increased even after intracardial injection of sterile sea water. When the latter, containing oils or coal tars, is injected into oysters a mass migration of phagocytes across the epithelial layers follows.

It is thus clear that molluscs contain powerful tissue-responding mechanisms which can protect them from invasion of ' foreign ' material. Why this mechanism does not operate against ' suitable ' trematode miracidia is not known, but presumably—as in other parasites—this phenomenon may be the result of a long period of association and adaptation before true parasitism began.

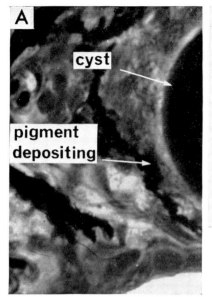

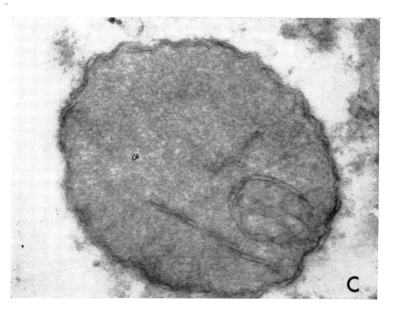

PLATE III. A. Metacercarial cyst of *Ascotyle branchialis* in the gills of *Rana esculenta*, showing appearance of melanin in host tissues (after Timon-David, 1961). B. Cercaria of *Posthodiplostomum cuticola* in an active swimming phase with tail thrust forward; the arrow indicates the direction of motion. Electronic flash photograph, 1/300 sec. (after Dönges, 1964). C. Mitochondrion from the parenchyma of *Fasciola hepatica*. Note that not more than one or two cristae are present and that a complex circular membrane is present. × 50,000. (after Björkmann and Thorsell, 1962)

Trematodes

Insects

The reactions of insects to the presence of parasites has been reviewed in detail.[274]   Cercariae normally invade aquatic insects through the intersegmental membranes, the gills, or the alimentary canal.   The epidermis reacts rapidly, often becoming thick and brown, but the reaction is usually too late to prevent infection; pigmented spots may appear at the site of entry.   The gut wall does not appear to react to penetration by cercariae.

Within the haemocoele many cercariae encyst rapidly and become encapsulated by their host.   Encapsulation seems confined to those species which encyst, and it is thought that possibly a substance involved in metacercarial cyst formation may be involved in invoking encapsulation on behalf of the host.

A capsule is formed by the haemocytes which accumulate about an invading parasite and flatten themselves over its surface.   The capsule grows by the addition of further haemocytes, the outer ones forming closely applied ' spindle ' shaped cells.   With time, the inner layer of the capsule degenerates as a cellular structure and becomes somewhat translucent.

Crustacea

Crustacea appear to react to trematode larvae in much the same way as insects, forming a capsule of connective tissue.

Platyhelminths

Trematode metacercariae may occasionally be found in planarians and even in cestodes.   There is no evidence that the wandering amoeboid cells which occur in this phylum react in any way to the presence of trematode larvae.   There is some evidence, however, that in trematodes they are phagocytic[45] (p. 183).

II: Physiology of the Host-Parasite Relationship.

II. Immunological Reactions to Trematodes: General Principles

Basic Concepts of Immunology

Antigens and antibodies

Immunology is perhaps one of the most rapidly expanding fields in parasitology, and it has become increasingly important for biologists to understand the basic principles of this subject. General accounts are given in several texts.[139, 296] Only the basic principles of immunology will be briefly summarized here.

Immunity (resistance) is essentially a reaction of the whole body in attempting to dispose of foreign material. The basis of resistance (see p. 159) to helminth infections is essentially the same as that of resistance to viruses, bacteria and non-living material. It is thought to depend on the fact that the host is able, in some way, to recognize that 'non-self' material is present in its tissues. 'Non-self' material is defined as being that which has a molecular configuration in some way different from that of the biological material normally present in the bloodstream or tissues of the host.

Such non-self material is said to be *antigenic*, and the molecules it releases or exposes to the host are termed *antigens*. Helminth antigens are sometimes divided into two groups: *somatic* antigens and *metabolic* antigens.

Somatic antigens are believed to take the form of the molecules released from exposed body surfaces such as the covering of an egg or the tegument of an adult trematode. *Metabolic* antigens are stated as being metabolic products, such as excretory or secretory products, which are released from a parasite into the host tissues or body fluids. It is not yet possible to say clearly that this distinction is justifiable.

Antigens are macromolecules of protein or complex polysaccharides; the rôle of lipids as antigens is still undefined. In order to act as an antigen, a molecule must have a minimal molecular weight of about 10,000 and its surface configuration must have a repetition of certain chemical groupings.

Presence of antigen in a host provokes the appearance of a soluble protein entity, termed *antibody*, in the blood, tissue fluids and sometimes the cerebrospinal fluid and certain cells. A specific antibody has the property of combining with a specific antigen either at the several different sites which provoked its initial production or, occasionally, with one having an identical antigen site. Antibody has been primarily identified as a complex group of heterogeneous but structurally related proteins known collectively as the immune globulins[96]; other globulin fractions may occasionally be involved. The actual mechanism of production of the antibodies is still a matter of controversy, although the ' clonal selection theory ' of Burnet[27] appears, at present, to be the most acceptable hypothesis put forward.

Antigen-antibody reactions

When antibody reacts against antigen it may invoke clumping of antigens—or clumping of cells bearing antigenic surfaces such as foreign blood cells—or it may cause lysis of antigen-containing cells—or it may provoke a number of other reactions listed below. Many of these reactions may be carried out *in vitro*.

The proportions in which antigens and antibodies are present has a marked effect on the resultant precipitates found *in vitro*. When the amount of antigen is equivalent to, or in excess of the antibody, precipitates are formed (fig. 52), but when excess antibody is present, there is not enough antigen to satisfy the bivalency of antibody, so that a ' lattice work ' is not formed and hence no precipitation takes place.

Serum of an animal which has been subjected to invasion of its blood or tissues by a parasite and which contains antibodies, is termed *antiserum* (or sometimes, less aptly, *immune* serum); the resulting type of immunity is termed *active immunity*. Immunity can also be introduced into a non-immune animal by injection of antiserum and the type of immunity produced in this way is termed *passive* immunity, since the recipient animal has not itself taken part in the establishment of the immune state; many human

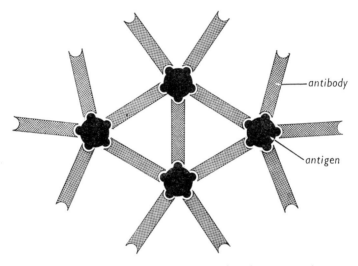

FIG. 52.   Diagram showing the interactions between antigens and antibodies. (after Smyth, 1962)

immunization procedures come into this latter category. Passive immunity is generally short lasting, as opposed to long lasting active immunity.

Routine serological tests

There are a number of properties of antigen-antibody systems which have been developed as useful laboratory methods of testing the suspected presence of antibody or antigen and measuring it quantitatively. Antibody concentration is expressed by the *titre* of a serum (i.e. the number of antibody units per unit volume of undiluted serum). The technical details of these methods are beyond the scope of this book, but are dealt with elsewhere.[152, 351] The most common methods widely used in serology are outlined below:

(a) *Precipitation.* With suitable concentrations of antigen and antibody, the aggregation of soluble antigen by antibody action may be observed *in vitro*. The solution becomes turbid and an insoluble antigen-antibody complex precipitates out.

(b) *Agglutination.* This phenomenon is essentially the same as that of precipitation, except that the antigens are larger and either particulate or cellular (e.g. bacteria or blood cells). Agglutination of red blood cells is referred to by a special term, *haemagglutination*.

A modified agglutination method much used is to use inert particles of polystyrene, latex, collodion, bentonite or red cells (either untreated or treated with tannic acid), as ' carriers ' and coat them with a variety of antigens. These can then be used for detection of antibody as shown by agglutination of the coated particles. The haemagglutination techniques using red cells are the most sensitive and can detect as little as 0·003 μgm. of antibody measured as antibody nitrogen.

(c) *Lysis.* Under certain conditions antibody may lyse cells acting as antigens. The phenomenon is limited to erythrocytes (*haemolysis*), a few bacteria and protozoa. The mechanism of lysis is not understood, as it involves another component of serum, termed *complement* (see below), which has the property of lysing cells after they have been ' sensitized ' by antibody.

(d) *Complement fixation.* The nature of complement is not clear. Complement, which has been defined simply as ' cytotoxic activity of serum ', comprises at least six factors or components which act in a definite sequence to damage the cell. The series of reactions which comprise the complement system is apparently initiated by antibody. Complement is present in greatly varying proportions in sera of different animals, but the level is constant for each species. Guinea-pig serum is widely used for experimental work, as it contains more complement than most other sera.

Complement is actively ' absorbed ' (=bound to) by antigen-antibody complexes in serum, and this property may be utilized as a delicate method for detecting antigen-antibody reactions. It finds its greatest application in serology in the Wasserman test for syphilis which may be quoted as an example.

The serum, suspected of containing antibodies to the particular parasite in question, is mixed with standard parasite antigen and added to fresh guinea-pig serum (as a complement source). After allowing the mixture to react, ' sensitized ' sheep red cells are added. If the cells lyse, the level of free complement has remained high, and the serum is negative for parasite antibody. On the other hand, if the cells *fail* to lyse, the complement has been bound by the antigen-antibody complex and the test serum is positive for antibodies to the parasite.

(e) *Hypersensitivity and intradermal reactions.* Hypersensitivity is a peculiar antigen-antibody reaction characterized by a heightened response to invasion by foreign materials. Thus, a guinea-pig injected with foreign protein (e.g. ovalbumin) becomes ' sensitized ' to that protein. If a further large challenging dose of antigen be given, the body suffers a violent general systemic shock (an anaphylactic shock), which may be fatal. This is in part due to the effect of a special type of antibody called a *reagin* or skin-sensitizing antibody and in part to the release of pharmacologically active substances—notably histamine, serotonin and bradykinin. The source of much of the histamine is apparently the mast cell.[58] By restriction of the antigen to a small area of skin, *local* hypersensitivity may be produced. These *intradermal* reactions are useful in the detection of certain helminth diseases, the test being carried out by injecting an extract of the parasite or placing some powdered parasite on a small area of previously scarified skin. A positive reaction, which essentially tests for the presence of a reagin, is indicated if a weal develops in the test area within 10–20 minutes.

Immunodiffusion

General principles. In addition to modifications and improvements of the more routine methods mentioned, new methods of ever-increasing discrimination for the detection of antigens and antibodies are becoming available. Most of these are related to the introduction of refined precipitin methods involving the use of gels and other support media, the general principle of the method being termed *immunodiffusion*. Detailed techniques are given by Crowle[57] and W.H.O.,[351] and the application of the method to helminth material in particular has been reviewed by Kent.[160, 161]

The principle of the method is that, when antigen and antibody are allowed to diffuse towards each other in agar, they form one or more separate precipitin bands according to the number of specific reactants present i.e. a complex mixture of antigens will produce a number of bands (fig. 53).

Double diffusion in tubes. The antibody and antigen are most conveniently mixed with agar (usually 0.6 per cent.) and separated by a layer of pure agar, so that there are three parts to the system.

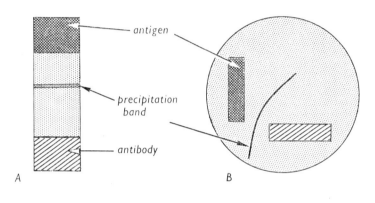

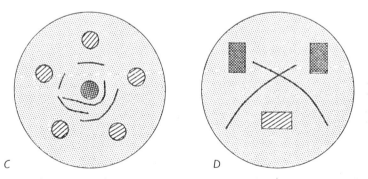

FIG. 53. Various forms of double diffusion-in-gel techniques. A. Tube technique of Oudin. B, C, D. Variations of the plate technique of Ouchterlony.

Precipitin bands form in the middle layer and, with strong anti-sera, may appear within a few hours (fig. 53A). This method is the most sensitive precipitin reaction available and can detect 0·3–0·9 μgm. of antibody nitrogen. It has its limitations, however; for the number of bands observed in agar tubes does not necessarily indicate the total number of systems present, as some of the bands in complex mixtures may be hidden by others. Such complex mixtures can be identified only by procedures of high resolving power, such as immuno-electrophoresis.

Double diffusion in plates. This is a useful technique for parasite antigens which may have complex systems, and is most used in the comparison of two antigen solutions with reference to one antiserum and vice versa. It depends on the diffusion of antibody and antigen towards each other from troughs or wells cut in agar. Such plates are known as Ouchterlony plates, after the inventor. If antigen and antibody (at equivalence) are allowed to diffuse from such troughs at right angles, a narrow line of precipitate is formed (fig. 53B). The tangent of the angle which this line makes with the antigen trough is equal to the square root of the ratio of the diffusion coefficients of antigen and antibody. When a rabbit antibody is used, the diffusion coefficient of the antigen can be worked out from the measured angle of the precipitate line and the known diffusion rate of rabbit gamma globulin. If it is required to examine more than one antibody, a series of cups can be cut out and the lines of precipitation compared (figs. 53C and 53D). Typical examples of the kind of result obtained are shown in fig. 55. Labelling antibodies with dyes or isotopes has been used to increase further the sensitivity of these methods.

Immuno-electrophoresis. This technique consists of the electrophoretic separation of a mixture of antigens carried out on a supporting medium such as agar, gelatin, or cellulose acetate paper. This is followed by the linear application of antiserum which is allowed to diffuse towards the separated components. Conversely, the antiserum may be separated by electrophoresis and the antigen applied linearly.

Fluorescent antibody techniques. These make use of the fact that proteins may be labelled with fluorochrome dyes, such as fluorescein isothiocyanate. These dyes have the property of absorbing ultra-violet light between 290–495 mμ and emitting a longer wavelength (525 mμ) green light. Thus, the protein labelled with this dye can be visualized by fluorescent microscopy. The cytological localization of labelled protein is excellent with this method, and material can be detected at a level as low as 1 μgm. protein per ml. body fluid. Tissue sectioning for the method must be carried out at low temperatures, and a microtome in a special cabinet at $-20°$ C. (cryostat) is used.

Serological Reactions in Specific Trematode Infections

General account

Investigations of the immunology to trematode infections has been almost entirely confined to several families of worms—the Schistosomatidae, the Troglotrematidae (especially *Paragonimus*) the Opisthorchiidae and the Fasciolidae, a position undoubtedly related to the availability and economic importance of these parasites. This does not necessarily imply that these species are the best experimental models for immunological studies.

Much experimental work has been concerned with the detection and estimation of antigen-antibody interactions and their use as diagnostic tools. A number of attempts to develop vaccines against helminth infections have also been made.

Antigen-antibody interactions

It is fairly easy to demonstrate that antibodies to trematode antigens develop in infected hosts. Intensive work on schistosomes has shown that all stages in the life cycle—egg, miracidium, cercaria and adult—possess antigenic properties. The term ' antigen ' is used in this text in the widest sense, i.e. any material stimulating antibody-production—bearing in mind that a number of antigens are almost certainly involved in each stage. These are often referred to by the general term *antigen mosaic*. The following examples, based almost exclusively on schistosomes, indicate the kind of general reactions which occur with each stage.

Egg

When a live schistosome egg is placed in antiserum, a precipitate in the form of small globules appears contiguous with the edge of the egg-shell (fig. 54B). Sometimes these appear as single globules; sometimes they form chains reaching a length of 150 μ or longer. The precipitate can be detected as early as two hours after incubation, but the effect is best seen after twenty-four hours at 37° C. Host serum apparently becomes positive about the fortieth day after infection at a time when eggs are produced (but see below). If drug treatment is given, egg production ceases and the serum once more fails to give a positive reaction.[19] The precipitate formed is made up of an antibody-antigen complex and the

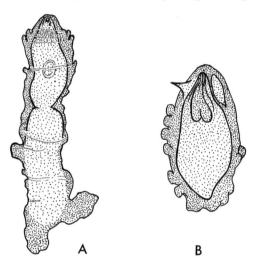

FIG. 54.   Immunological reactions of *Schistosoma mansoni* in antiserum. A. *Cercarienhüllenreaktion* (CHR).   B. Circumoval precipitin test (CPT).   (after Brumpt *et al.*, 1962)

test, which is known as the *circumoval precipitin (COP) test*, can be used for testing host serum.   It is of especial interest to note that serum from unisexual infections of schistosomes (p. 143) whether males or females (which do not produce eggs), give a weak COP reaction;[288] which suggests that antigenic specificity is quantitative rather than qualitative.   The antigen appears to be secreted or excreted by living worms, or alternatively it may be an unstable antigen present only in the living cells of schistosomes.[287]   The appearance of a precipitate around the egg does not give an indication of the complexity of the egg antigens;   for, when the antigen-antibody interaction is tested more precisely by immuno-diffusion techniques (such as the Ouchterlony plate), up to four precipitate lines have been identified. Eggs appear to share a number of antigens with cercariae (see fig. 55).

Miracidium

(*i*) *Somatic antigens*. When schistosome miracidia are placed in antiserum some become immobilized within five minutes.

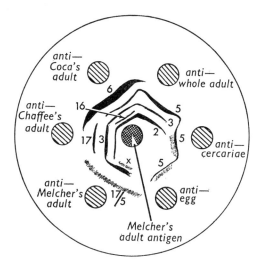

FIG. 55. Ouchterlony reaction of Melcher adult antigen of *Schistosoma mansoni* with various antisera prepared against life cycle stages of the parasite. (after Kagan and Norman, 1963)

This appears to be largely due to the absorption of antibody at the surface and the agglutination of the cilia. The phenomenon is termed ' miracidial immobilization '. The antigen(s) stimulating production of the miracidial immobilization reaction is/are present in all mammalian stages of the parasite—eggs, cercariae and adults.[281]

(*ii*) ' *Metabolic* ' *antigens*. When schistosome eggs are allowed to hatch in a small volume of water (1000 eggs/ml.) and incubated for a short period, the metabolic products so produced are strongly antigenic and will produce a weal if injected intradermally into an infected host, or will produce a precipitate if incubated in a tube of antiserum.[284] Schistosome antigens have also been detected in the urine of patients with schistosomiasis.[287]

Cercaria

Two main serological phenomena occur when cercariae are placed in immune sera, the *Cercarienhüllenreaktion* (CHR)[343] and the *agglutination* reaction. Such serum also has a cercaricidal action, but this is often hidden by the CHR.[286]

(*i*) *Cercarienhüllenreaktion.* Briefly this assumes the form of an envelope—the *pericercarial envelope*—first described around the cercariae of schistosomes (fig. 54A). In more detail, the process takes place in *S. mansoni* as follows.[308] Within five minutes of immersion in antiserum, cercariae secrete orally large amounts of sticky material by which they become permanently attached to the substrate. During the first twenty minutes, the interspace between the outer and inner limits of the cuticle seems to thicken as though the cuticle were swelling or a fluid were collecting there. During the first hour, the cuticle continues to 'swell' until the cercariae appear to be ensheathed in tight transparent envelopes; during this period the envelopes lose their elasticity. Eventually the envelopes loosen and balloon away until cercariae are encased in wrinkled, loose-fitting structures somewhat resembling plastic cercarial moulds. The origin of the CHR is still somewhat obscure; in its final manifestation it results in a fully developed, hardened, transparent, permanent envelope, which is plainly visible even under low powers ($\times 30$) and appears to be protective to cercariae.

The materials forming the envelope are not completely heat-stable and their appearance in serum is not dependent on the sexual maturity of the parasite. It was first thought that the pericercarial envelope was a simple product of a cercaria-secreted antigen and a serum antibody. It now appears that a cercaria is enclosed in a closely fitting mucosal film (possibly, but not certainly, secreted by one of the 'cephalic' glands just before emergence) and it is the antigenic reaction of this film with serum antibody which produces the pericercarial envelope in host antiserum.[304, 305]   This view is strongly supported by the fact that after a cercaria has penetrated and become a schistosomula, it loses its ability to form pericercarial envelopes.   This further emphasizes the physiological differences between cercariae and schistosomulae (see p. 128).

Somewhat comparable results have been obtained with ex-cysted metacercariae of *Fasciola hepatica* in antisera; but in this case a mucous film is missing and the antibody-antigen precipitation is weak and takes the form of precipitates around the oral aperture.   The precipitates appear after sixteen hours' incubation and the antibodies are present in rabbit serum twelve days after infection.[353]   Comparable envelopes have also been found to occur around strigeid cercariae in antisera from chickens.[179]

(*ii*) *Agglutination*.   This is essentially a manifestation of the same phenomenon which is responsible for the CHR.   Under certain conditions, not clearly defined,[47, 308] cercariae become sticky in antiserum and will agglutinate if present in large numbers. Permanent cercarial agglutination seems to occur when the host serum is so handled that cercariae tend to settle out and do not form well-developed pericercarial envelopes.

Adult worms

A number of fractions of adult worms have been prepared and tested for antigenicity.   These are often given names such as *Melcher's* adult, *Coca's* adult etc. (fig. 55), after the person who developed the method of extraction.   Lipid and carbohydrate fractions have been found to be non-antigenic when tested with antisera by the complement fixation test.   As might be expected, the protein fractions yield a number of antigens, all of which appear to be present in the acid insoluble protein fraction. Some of the antigens prepared from other species, such as *Clonorchis sinensis*, give cross reactions with sera of individuals with paragonimiasis and schistosomiasis.[272]

Identification of trematode antigens

The introduction of agar-gel immunodiffusion methods (p. 174) has revolutionized the isolation and identification of helminth antigens.   Whereas early precipitation techniques indicated that a particular extract might contain one or two antigens, agar-gel methods invariably reveal that many more are present.

By the use of such methods, then, a total of seven specific adult worm antigens, three cercarial and five egg antigens have been identified in *S. mansoni*,[153] but a further 25 bands have not been identified. The principle of the method is evident from fig. 55. Each specific band identified is given a number, 1, 2, 3 etc., and it is often found that many antigens have bands in common. Thus, in fig. 55 the innermost band—No. 2—is common to the three anti-adult sera tested. Similarly the anti-cercaria and anti-egg sera share band 5 with each other and with 2 anti-adult sera. The method outlined above has been applied to a number of other helminths, including several trematodes. It is becoming clear that some helminths may share antigens with other species sometimes belonging to different phyla, and this may give rise to false positives in serological tests.

Thus, immunoelectrophoresis of antigens of *Fasciola hepatica* have revealed the existence of 15 antigens, only 5 of which are specific to this trematode. In this species, common antigens have been found with the trematodes *S. mansoni* and *Dicrocoelium dendriticum*, the cestodes *Taenia saginata* and *Echinococcus granulosus* and the nematodes *Onchocerca volvulus* and *Trichinella spiralis*.[11]

Serological diagnosis

Most of the routine serological methods discussed earlier, complement fixation, haemagglutination, agglutination—intradermal tests (p. 172)—have been used for the diagnosis of trematode infections such as schistosomiasis, fascioliasis or paragonimiasis. The major problem has been to obtain standard ' antigen ' when extracts or ground dried worms have been used as antigens. It is clear from the use of immunodiffusion methods that such material is complex and likely to give false positives with other infections. The circumoval test and methods involving live cercariae are also widely used in diagnosis. The limitations of most of these methods are now widely appreciated, but nevertheless, when critically used, they form useful diagnostic tools. Further work will undoubtedly increase their precision and reliability.

Phagocytosis in trematodes

It is interesting to note that trematode tissue itself may possess some innate defence mechanism at a cellular level. Thus, cells which are believed to represent primitive blood cells and termed ' lymphocytes ', ' haemocytoblasts ' or ' haemocytes ', have been reported in a number of species.[45] These cells are capable of phagocytizing foreign particles (such as indian ink particles) and eliminating them through the alimentary tract.[45] It is clear that such cells could play a rôle in the defence of trematode tissue against invasion by hyperparasites such as microsporidia; but this has not been actively demonstrated. In some species, the haemocytes are associated with the ' lymph ' (p. 18) or excretory systems.

12 : Physiology of the Host-Parasite Relationship

III. Immunological Reactions to Trematodes : Innate and Acquired Resistance

Innate Resistance

General considerations

If the infective stage of a parasite comes in contact with a potential host and fails to become established in it, the host is said to possess *innate* or *natural* resistance to that parasite. The mechanism preventing the establishment of a parasite within a host may often be complex and difficult to define in precise terms. The term ' resistance ' implies that an *active* process of defence (in what may be called a ' military ' sense) is taking place (p. 159). On the other hand, the term *insusceptibility* implies that adverse conditions of the behaviour, ecology or physiology of the host eliminate the possibility of the parasite in question becoming established in a particular host. For example, a parasite may be adapted to live in a host whose bile has a particular composition, and therefore cannot live successfully in a host whose bile differs markedly from that composition (p. 131). This could be termed ' physiological or biochemical insusceptibility '. Included under the same heading could be instances in which failure to develop was linked with some other ' unsuitable ' physiological condition, such as the body temperature or the local pH or pCO_2 in a particular tissue site. Again, the diet of the host may be lacking in a nutritional factor essential for the parasite's development.

Behavioural or ecological factors must often prevent a parasite from making contact with a particular host in which it would be theoretically possible for it to become established. Thus, it has been shown (p. 163) that many parasites (e.g. *Fasciola hepatica, Cryptocotyle lingua, Schistosoma mansoni*) may develop in a wide spectrum of hosts with which they may not normally come in contact—a fact which, perhaps, has been fully appreciated only within the last decade. In dealing with a trematode, as with other parasite species, it is customary to speak of the ' natural ' or ' normal ' host, i.e. the host in which it is more usually found in the field. This does not eliminate the possibility (however remote) of the existence of another host, perhaps living in some obscure part of the world, which could serve as a ' better ' host—if we use the latter expression to imply that greater protein synthesis, in terms of egg production, could occur.

Innate resistance to schistosomes

Innate resistance has probably been more thoroughly studied in schistosomes than in any other group; it appears to vary considerably with species.[186, 305] *S. japonicum* encounters little natural resistance among vertebrates, either to initial invasion or to full development of egg-producing adults. Many vertebrate hosts thus act as ' reservoir ' hosts for this species; these include dogs, cats, oxen, pigs, sheep, goats, horses, water buffalo and various field rodents. Similarly, *S. mansoni* has been reported as a natural infection in 20 species of vertebrates, and some 54 other vertebrates have been infected experimentally.[305] As was shown earlier (p. 163), it is possible to draw up a table (table 27) of a graded series of host-resistance to *S. mansoni* ranging from none (in which normal eggs are produced) to almost complete resistance —in which sexual maturity is never achieved.

Innate resistance is further complicated by the fact that different ' strains ' of host may show different degrees of resistance and, similarly, different ' strains ' of parasite may show different powers in their ability to infect.

In contrast, in the case of *S. haemotobium* only two naturally infected hosts other than man have been reliably reported, although some dozen species have been experimentally infected.

Many laboratory animals, albino rat, guinea-pig, rabbit, cat and dog, show complete innate resistance. Comparable results have been described for species of schistosomes, other than those that attack man.

The horny layer of the skin, which is the first tissue with which schistosome cercariae come in contact, is probably chiefly involved in natural resistance by leucocytic destruction of the schistosomulae as they move through the stratum corneum. The basement membrane and the intercellular ground substance of the skin may also represent a formidable barrier, especially in ageing hosts. Schistosomulae which do penetrate these sites may be destroyed later in the lungs or liver.

Innate resistance in other species

The pattern described above for schistosomes is generally found with other species, that is a grading of resistance in different host species. Some trematodes have a very wide host spectrum and can even develop in hosts from different phyla. For example, *Cryptocotyle lingua*, commonly an intestinal parasite of gulls, occurs naturally in a wide range of other bird species, but at least one species of duck exhibits innate resistance. *C. lingua* can also develop in rats, cats and guinea-pigs.

Acquired Resistance

General considerations

Acquired resistance to trematodes is a manifestation of a general immunological response to invasion by a parasite. It may be acquired in a number of ways, which fall into two main groups—actively and passively.

A. Actively:
  (*a*) naturally, as a result of a natural infection with either
      (*i*) a parasite of the same species (homologous)
    or (*ii*) a parasite of a different species (heterologous)
  (*b*) artificially, after injection of
      (*i*) 'somatic' vaccines—extracts or homogenates of trematode tissues,
      (*ii*) 'metabolic' vaccines—metabolites of trematodes,

(*iii*) ' live ' vaccines—living adult or larval trematodes attenuated (weakened) by treatments such as irradiation.

B. *Passively*—by transfer of antibodies in injected antiserum.

One fundamental point must be stressed at the outset in discussing acquired resistance, namely, that the appearance of an antibody in host serum does not necessarily imply that the resistance of the host has been increased in any way. It will be increased only if the antibody acts on the parasite in such a way as to interfere with its general growth or metabolic processes or to upset its relation to the host tissue by enclosing it in a precipitate or by blocking its alimentary, excretory or reproductive orifices.

Thus, the demonstration of an antibody in host serum implies only that the parasite has made a sufficiently intimate contact with host tissue for the latter to recognize it as foreign and produce antibodies accordingly.

As is shown below, many attempts to develop suitable vaccines against trematode parasites have been made but few of these have stimulated production of antibodies which attack the vital functions of a parasite to any extent. Antibodies which can be shown to confer a measure of protection on a host are known as *protective* antibodies and the antigens which stimulate their production are sometimes termed *functional* or *essential* antigens, the latter term being used in the sense we use when we speak of ' essential ' amino acids. The antigens are essential to the parasite, and if the processes in which they are concerned are attacked or modified in some way by antibody, the parasite can no longer survive.

Actively acquired immunity

Much of the evidence in this field is based on laboratory infected animals, but there is a certain amount of zooetiological data available, especially on schistosomes. Comprehensive reviews of this field are available.[270, 305]

Schistosomes

(*i*) *In natural hosts.* There seems to be no doubt that some degree of acquired immunity to challenging infections is developed after previous infections with schistosomes; this immunity appears to be of both humoral and cellular origin. There are numerous examples of human communities in different areas of the world which show complete or partial resistance to invasion of schistosomes. The earliest observation was made in 1916 by Fujinami who noted that inhabitants of endemic districts in Japan were generally not seriously ill with schistosomiasis, but uninfected workers or cattle who settled in these areas became sick and often died shortly after their first exposure. However, those who recovered appeared to become more resistant to reinfection. There are also many examples from experimental infections. One of the difficulties in this type of work is that of finding suitable ' markers ' by means of which the degree of resistance is measured. Clearly, failure to develop to maturity is a fairly clear-cut criterion, but the number of eggs produced has also been used, as has the ability of the host to survive an otherwise lethal infection.

(ii) *In experimental hosts.* Many experiments have been carried out to test the degree of resistance of schistosome-infected hosts to challenging infections. These results have shown a wide variation; many workers have demonstrated a substantial degree of protection; somewhat fewer have reported negative results.

These conflicting results can be exemplified by reference to experiments with the rodent parasite, *Schistosomatium douthitti.* Thus, mice and rhesus monkeys, after single or multiple immunizing infections, were found in some experiments to have fewer adult worms than those previously uninfected, the reduction varying from 0 to 50 per cent. in different experiments.[305] In addition, the inflammatory reaction of the skin to cercarial penetration was accelerated, intensified and more localized in immunized than in non-immunized mice. In contrast, other workers[143] with the same species could not demonstrate any reduction in numbers of adult *S. douthitti* after five immunizing and one challenge dose. Variable results have similarly been obtained in comparable experiments with other species of schistosomes, especially *S. mansoni* and *S. japonicum.* In *S. mansoni* increased local resistance to a subsequent challenge is developed as early as one hour after exposure (fig. 56). Much early experimental work may be open to criticism on the grounds that sufficient time was not allowed between the initial and the challenge infection for immunity to develop. There is some evidence that the production of eggs over a period may be necessary for the development of immunity, and it has been claimed that an *effective* degree of immunity cannot be detected until sixty days after the initial infection;[140, 141] antibodies can, however, be detected even in egg-less infections (p. 178). If confirmed, this result would invalidate much early experimental

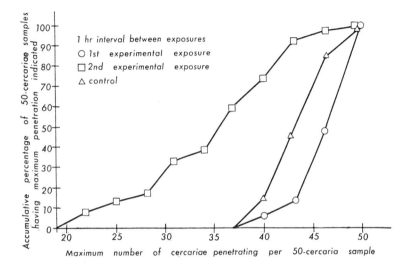

FIG. 56. Development of local resistance in mice to entry by challenging cercariae of *Schistosoma mansoni*. Accumulative frequency distribution showing percentage of 50-cercariae samples in which the maximum number penetrating was ≦ that indicated. (after Stirewalt, 1953)

work, but other workers have been unable to confirm this effect.[288] In other species, such as *S. japonicum*, immunity appears to develop at an earlier stage in the infection,[270] both cellular and humoral factors operating.[185]

It is interesting to note that immunization with heterologous parasites may in some cases give a better degree of protection than with the homologous parasite. This has been most elegantly demonstrated in reciprocal experiments with *S. mansoni* and *Schistosomatium douthitti*.[143] Mice immunized by five exposures to 10 cercariae of *S. mansoni* and challenged with 50 *S. douthitti* cercariae were found to show a highly significant reduction in infection in comparison with the challenge control. In contrast, mice infected with *S. douthitti* in the same way and challenged with *S. mansoni*, showed little evidence of such reduction. This remarkable result is of particular interest; for, although reciprocal cross reactions have been demonstrated with other organisms (the classical case in the field of virology is, of course, immunization against small-pox by cow-pox virus), it is generally observed that the strongest immunological response occurs with the homologous antigen.

As emphasized earlier, however (p. 187), the presence of humoral antibodies does not necessarily confer protection on the host; other complex factors, such as host tissue responses, are involved. There is reason for believing that the mouse is a ' better ' host for *S. douthitti* than for *S. mansoni*. If this is so, it could be concluded that the less well adapted *S. mansoni* provoked in the mouse a response of sufficient magnitude to affect seriously the development of *S. douthitti* when it subsequently invaded the host. Conversely, the better adapted *S. douthitti* did not elicit a response that appreciably affected later invasion by *S. mansoni*. In both cases the host response was not sufficient to influence subsequent infections with the homologous parasite. In this case, then, it is clear that *S. douthitti* is particularly sensitive to the host reaction elicited by *S. mansoni* (but not vice versa).

Experiments have also been carried out using cross infection between ' strains ' of the same species—such as the ' Japanese ' and ' Formosan ' strains of *S. japonicum*—and some reciprocal protection has been demonstrated.[270]

Other species

The acquired resistance of species other than schistosomes has been poorly studied. In infections by *Fasciola hepatica*, although circulating antibodies may be readily detected, a significant degree of resistance to subsequent challenge does not develop. The position is somewhat similar with *Clonorchis sinensis* and *Paragonimus westermani*, and a small level of resistance has been found by some workers. Antibodies can readily be detected in the serum of hosts infected with these trematodes and a number of diagnostic tests, of which the intradermal ones are probably the most reliable, have been developed.

Artificially acquired immunity (vaccination)

General considerations. Probably few fields in parasitology have aroused such interest within recent years as those concerned with attempts to develop vaccines against diseases due to metazoan parasites. A considerable degree of success has been claimed for vaccines against nematode diseases of domestic animals;[294, 295] but only modest success has been achieved with trematode infections.

Various methods have been used to prepare antigenic material as vaccines; these fall under three headings:

(*i*) whole homogenized fresh or dried adults, eggs or larval stages or extracts of these or formalin-killed material of these same stages,

(*ii*) metabolic products from adults or larvae,

(*iii*) larval stages attenuated by irradiation.

Vaccination with homogenates or extracts. There is no concrete evidence that dead trematode material, whatever its source—adult, larva or egg—can provoke any marked degree of resistance to subsequent infections. The results obtained have varied considerably with species and strain of parasite and host, and especially with the method of preparation of the antigens in question, and with the experimental protocols adopted in different laboratories. The handling of homogenates to prevent denaturation of proteins may be one of the reasons for the variation in the reported results. For example, although the mouse is not protected by vaccination with cold-stored homogenates of *S. mansoni*, this vaccine gives substantial protection to rats. Since so many variables are involved, standardization of antigens is difficult. A large literature exists in this field, the results of which have been critically assessed and reviewed.[270, 287, 305] Although some workers maintain that with schistosomes there occurs a reduction in worm size, degree of development, or a delay in the death of the host, other workers maintain that no protection was obtained by this vaccination technique. In one group of experiments, the percentage recovery of worms from immunized rats was 4·6 per cent. and 5·4 per cent., compared with 9 per cent. in the controls, and in another group of experiments it was 0·2 per cent. and 0·3 per cent. compared with 1·4 per cent. in the controls.

Vaccination with metabolic products. A limited number of attempts to vaccinate against trematodes by means of collected metabolic products have been made. The technique involves the culture of adults, larvae, miracidia or eggs in saline or other suitable maintenance media for a short time and then using the

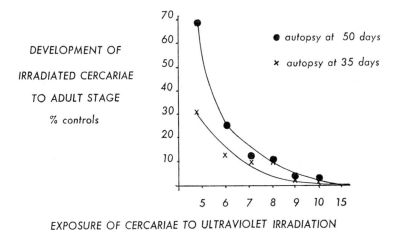

DEVELOPMENT OF

IRRADIATED CERCARIAE

TO ADULT STAGE

% controls

● autopsy at 50 days

✗ autopsy at 35 days

EXPOSURE OF CERCARIAE TO ULTRAVIOLET IRRADIATION
seconds

FIG. 57. Influence of ultra-violet irradiation on the cercariae of *Schistosoma mansoni*, as judged by ability to reach maturity. (after Standen and Fuller, 1959)

medium with the excreted or secreted metabolic products as a vaccine. Results with *S. mansoni* and *S. japonicum* have been the same as those reported in the previous section, i.e. a barely significant degree of resistance has been reported in some cases.[182, 271] Results in general, have not been convincing.

Vaccination using irradiated materials. The fact that some degree of resistance is developed in naturally infected hosts has led to the conclusion that the most ' protective ' antibodies are induced by *living* worms. Attempts, therefore, have been made to use living worms attenuated in some way so that they are unable to reach full maturity in the host. Ultra-violet light can be used for attenuation, but its use is difficult to control due to the short exposures (15 sec.) required to kill larvae[299] (fig. 57). Attenuation is better carried out by irradiation with X-rays or gamma radiation from sources such as Cobalt 60, a process first used on parasitic nematodes. Larval helminths treated in this way can undergo their normal migration in the host but only a very small percentage reach maturity. By this means the living—but ' attenuated '—

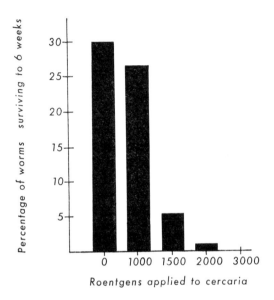

FIG. 58. Percentage of cercariae surviving to six weeks in hamsters after X-irradiation. (after Smithers, 1962)

larvae survive long enough to make contact with the host tissue but not long enough to reach maturity or to damage the host.

When applied to trematodes, such methods have had only limited success. As in earlier studies, most work has been carried out on schistosomes. The initial problem was to determine the radiation dose which would prevent cercariae from maturing but still allow them to migrate through the lungs to the liver. The effect of increasing doses of X-radiation on *S. mansoni*[287] is clearly shown in fig. 58. A dose of 3000 roentgens prevented any cercariae from surviving more than six weeks; 2000 r. prevented all but 1 per cent. of the worms from surviving beyond this time; levels of 1000 r. or less do not permit more than 25 per cent. survival. Worms which survive irradiation of more than 2000 r. treatment are retarded in development and unable to produce eggs. Cercariae which have been treated with 2000–3000 r. apparently pass through the lungs uninhibited, but after reaching the liver die within 2–4 weeks after initial infection. Dead flukes are

TABLE 29

Exposure of Rhesus monkeys to X-irradiated cercariae of S. mansoni
and result of homologous challenge
(Data from Smithers[287])

| | Immunizing cercariae | | | | | | Post-mortem results | |
Monkey No.	X-ray dose (r.)	No. in first exposure	Interval (wks.)	No. in second exposure	Interval (wks.)	No. of cercariae in challenge	Worm recovery	Lesions
39	—	—	—	—	—	1,000	328	Severe
44	—	—	—	—	—	1,000	586	Severe
46	—	—	—	—	—	1,000	420	Severe
37	2,000	3,000	6	10,000	8	1,000	166	Mild
38	2,000	3,000	6	10,000	8	1,000	341	Mild
45	2,000	3,000	6	10,000	8	1,000	456	Mild
40	3,000	3,000	7	7,000	6	1,000	73	Mild
47	3,000	3,000	7	10,000	6	1,000	207	Moderate
48	3,000	3,000	—	—	13	1,000	440	Severe

rapidly surrounded by intense cellular reactions consisting mainly of mononuclear cells with a few polymorphonuclear cells.

Using irradiated cercariae of various schistosome species as live vaccines, a number of attempts to vaccinate mice or monkeys against subsequent challenge infections have been made.[134, 270, 287] A typical list is given in table 29, and it shows the kind of conflicting results which may arise in this type of experiment. It shows that monkeys exposed to cercariae irradiated at 2000 r. developed a good partial resistance to a challenge infection, although their worm burdens were higher than anticipated. In monkeys exposed to one or two doses of cercariae irradiated with 3000 r. results were equivocal in that one animal (No. 40, table 29) developed a clear partial resistance, another (No. 47, table 29) contained 207 worms and a third (No. 48, table 29) showed no increase in resistance in comparison with the controls. Some success in vaccination has been achieved by using X-rayed cercariae of *S. japonicum* on monkeys.[134]

Comparable studies have been carried out with *Fasciola hepatica*[137, 138, 317] using X-irradiated metacercariae, again with conflicting results; partial immunity being obtained in some experiments and not in others. This species appears to be able to stand high levels of X-irradiation; for a small percentage of adults reached maturity and produced viable eggs even after doses of 8000 r. Doses of 12,000 r. or more proved fatal.[354]

Successful vaccination with larvae, attenuated by means of irradiation, may prove to be possible; but it is clear that much more exploratory work needs to be done in this field.

Vaccination by passive transfer. ' Passive ' transfer involves the injection of ' immune ' serum from an infected host to a non-immune host. Experiments with various species of schistosomes[182, 218, 271] have failed to demonstrate that a significant degree of resistance can be induced by this procedure.

SECTION II
PHYSIOLOGY OF THE ASPIDOGASTREA AND MONOGENEA

13: Physiological Problems in the Aspidogastrea and the Monogenea

General Considerations

The bulk of experimental work on trematode physiology has been carried out on the Digenea as discussed in earlier chapters. There remain to be considered two other groups traditionally classified with the Trematoda—the Aspidogastrea and the Monogenea. As indicated earlier (p. xv), modern work on larval forms[29, 191, 195] suggests that the latter group is phylogenetically nearer the gyrocotylidean cestodes than the digenetic trematodes, and recent classifications have placed them in a separate class of the Platyhelminthes, i.e. the Monogenea. Nevertheless, the possession of an alimentary canal and well-developed adhesive organs, as well as their occurrence in ' tissue ' rather than ' intestinal ' sites, suggest that the physiology of this group may more nearly approach that of the digenetic trematodes than the cestodes. For these reasons, the Monogenea are considered briefly here.

So little physiological work has been carried out on the Aspidogastrea and the Monogenea that knowledge of their general physiological processes is completely lacking in the former group and fragmentary in the latter. It is thus not possible to present anything approaching a coherent picture of the physiology of these groups; the most that can be done is to survey the isolated pieces of work which have been done, point to areas where differences from the Digenea are likely to occur, and indicate some of the more outstanding physiological problems.

Aspidogastrea

The Aspidogastrea are a group, containing nine genera, which are parasites of poikilothermic animals; chiefly molluscs and crustacea are infected, but fishes and chelonians also can act as hosts. Their most striking anatomical feature is an enormous adhesive apparatus which occupies almost the entire ventral surface. For a general morphological account of the group, see Baer and Joyeux.[7]

The best-known form is probably *Aspidogaster conchicola* which occurs in the pericardium of freshwater mussels of the genera *Anodonta* and *Unio*. The life cycle pattern is not clear. Eggs hatch in the definitive host and either continue their development there or escape to infect other molluscs. If infected molluscs are ingested by a fish host, the adult *Aspidogaster* can become reattached in the alimentary canal. Species of turtles, such as *Pseudemys*, have also been infected experimentally by feeding them with adult *Aspidogaster* via a stomach tube. This interesting group thus provides evidence of one way in which alternation of hosts within the Trematoda could have been achieved.

Nothing appears to be known of the chemical composition, digestion, nutrition, carbohydrate, protein or fat metabolisms or respiration of the adults, larvae or eggs. The nature of the egg-shell material, the mechanism of hatching and the physiology of reproduction are likewise unknown. It is possible that the enormous ventral sucker may prove to be a glandular organ and act as an organ for extra-cellular digestion as in the strigeid trematodes (p. 25).

This group may lend itself to physiological work, for specimens may be ' maintained ' (see p. 147) for long periods *in vitro*, although no data on the metabolic or cytological conditions of worms during and after culture are available. *Aspidogaster conchicola* has been ' maintained ' in the following media at 2–9° C.: 0·75 per cent. NaCl—38 days; mussel Ringer—39 days; Hédon-Fleig—29–38 days; mussel blood—75 days.[330] Clearly, the Aspidogastrea are a group open to much further physiological study.

Monogenea

Although a little more information is available on the physiology of the Monogenea, our knowledge of it is nevertheless slight.

General account

Monogenetic trematodes differ from digenetic trematodes in having a direct life cycle (i.e. with no alternation of hosts). This implies that there is a much closer ecological relationship between the egg, larval stages and definitive host than in the case of Digenea, which utilize an intermediate host or hosts. For example, in at least one instance (*Polystoma integerrimum*) the maturation of the genitalia in host and parasite appears to be controlled by the same hormonal system, with the result that both host and parasite larvae make their appearance simultaneously.[221] Host specificity in the group is well defined and appears generally related to the morphological adaptations developed by specific parasites for specific host sites, especially the gills (p. 201). Physiological tolerance to factors such as speed of current, presence of suspended matter, temperature, oxygen tension and chloride ion content is also important in the ecology of gill monogeneans;[247, 248] some species show a substantial degree of tolerance to changes in salinity.

Lack of an intermediate host also implies that the means of location of, and subsequent attachment to, the definitive host must be highly developed in the group, a condition reflected by the morphology of the larval forms.

Monogeneans are chiefly parasites of fish, amphibia and reptiles, but have also been reported from parasitic isopods, cephalopods and aquatic mammals. They are chiefly ectoparasitic forms, being found on the gills, the gill chamber, skin, buccal cavity and other superficial locations, but are also not uncommon in the uterus and body cavity. Aberrant locations are the excretory system (*Acolpenteron petruschewskyi*), and the heart (*Amphibdella torpedinis*).[269] The bloodstream forms of the latter species are probably juvenile forms.

The morphology of the Monogenea has been covered in a number of texts[7, 29] and will not be discussed in any detail here.

Adhesive attitudes

The adhesive attitudes and structure and mode of operation of the organs of attachment have been the subject of considerable study, especially by Llewellyn and his co-workers.[158, 188-193, 197] The chief organ is a posterior disc which bears hooks or suckers, or both. There is also an anterior adhesive organ, generally in the form of an oral sucker, often absent or poorly developed. The terms *opisthaptor* and *prohaptor* have been used for these structures by some authors. As these may be *mixed* organs, part being used for feeding and part for attachment, it is difficult to define them by precise functional terms.

Most of the studies have been made on gill and skin parasites, particularly the former. In general, and probably without exception, gill forms attach with their posterior adhesive organs near the gill arch of the host with the anterior end nearer to the distal end of the primary lamellae.[188] This results in the attached ends of the worms lying upstream relative to the current passing over the gill, with the mouth of the parasite lying downstream. Worms also show a tendency to attach to the secondary gill lamellae rather than the primary. Apart from the general features, monogeneans exhibit considerable variation in their adhesive attitudes, e.g. attachment to one or more primary lamellae or to a particular hemibranch, or establishment between primary lamellae of the same or different hemibranchs. The adhesive attitude adopted by any particular species is comparatively consistent, and the morphology of a worm, with respect to width of body, development of peduncles on opisthaptor, and asymmetry, can be related in general to that attitude. The simplest method of adhesion is for a worm to apply its whole body to the secondary lamellae of one surface of a primary lamella. This method is found in the European genera: *Octostoma, Mazocraës, Plectanocotyle, Discocotyle* and *Microcotyle*.

In gill forms, the opisthaptor is generally adapted for clamping, gripping or sucking, and the number of adhesive organs may sometimes be increased (e.g. *Axine belones*) from the more usual 6 or 8 to a number varying between 50 and 70. The clamps work on an extrinsic muscle-tendon-fair-lead-hinged-jaws system worked out in detail for several species, e.g. *Discocotyle sagittata*.[197] In some species, e.g. the *Diclidophoridae*, the clamping action of jaws is brought about indirectly by suction pressure of a diaphragm.[192] Hooks may also be present and used for assisting attachment. Thus, *Octostoma scombri*, which uses four pairs of clamps to grasp one or two secondary lamellae, also possesses a pair of hooks with powerful muscles, which perforate three or four secondary lamellae.[189] It is interesting to note that the hook-like structures found in larval forms contain cystine sulphur, as in the eucestodes and cestodarians, thus strengthening the conclusion that cestodes and monogeneans are closely related (p. 198).[199]

Many species, e.g. *Anthocotyle merluccii*, *Axine belones*, *Pseudaxine trachuri*, *Gastrocotyle trachuri*, show asymmetrical development of the adhesive organs—a modification which brings the longitudinal axis of a worm to lie parallel to the gill-ventilating current. Some workers have claimed that parasites are capable of moving from one site on a gill to another after the initial attachment; but the evidence for this is not unequivocal. The mechanics of the attachment organs have been analyzed in detail in a number of species, e.g. the family Diclidophoridae,[192] *Discocotyle sagittata*,[197] *Octostoma scombri*,[189] *Plectanocotyle gurnardi*,[188] *Gastrocotyle trachuri*.[193]

In species which attach to smooth surfaces, such as the wall of the bladder or the skin, suctorial action is generally used. In *Entobdella soleae*, on the skin of the common sole, suctorial pressure is produced by the action of a pair of extrinsic muscles which lift a pair of girder-like hamuli embedded in the roof of the haptor.[158]

Feeding and digestion

General comment. In general, it can be said that members of the two sub-orders differ widely in their food and feeding habits. The Monopisthocotylea, which are mainly skin and gill parasites, feed almost exclusively on epidermis and the mucus secretions frequently associated with it. The Polyopisthocotylea, which are largely gill parasites, feed chiefly on blood, in some cases ingesting some tissue and mucus in addition.

These conclusions are based on a number of studies.[128a, 148, 157, 187, 329] Kearn[157] has given a general review of the problems of monogenean feeding and has stressed that, regarded as a food material, the fish epidermis has a number of valuable properties. Firstly, it is easily available and vulnerable to attack by the parasite; secondly, it lacks external keratinization; and thirdly, it possesses very considerable powers of regeneration after damage— for example, the epidermis of a loach (*Misgurnus fossilis*) can regenerate an area of 25–100 sq. mm. within 24–36 hours. That some monopisthocotyleans, such as *Entobdella soleae*, move from place to place is evidenced by the number of feeding wounds in the vicinity of individuals.[157]

Monopisthocotylea. In this sub-order, the feeding apparatus differs considerably from that of the blood-feeding Polyopisthocotylea. In the latter, the mouth is terminal, whereas in the monopisthocotyleans the mouth is generally not terminal but is ventrally situated. The morphology and mode of action of the feeding apparatus of *E. soleae* has been described in some detail.[157] The main part of the apparatus consists of a glandular ' pharynx ', better termed a feeding organ. As in many species, part of this organ is capable of protrusion and of becoming pressed closely to the epidermis, where it remains undergoing feeding for about five minutes. The organ contains about 60 gland cells and 45–49 papillae associated with it, the latter being arranged in a circle. It has been demonstrated experimentally that the glandular secretion is proteolytic in nature and capable of digesting the circle of epidermis enclosed by the organ. Elimination of undigested gut contents takes place by egestion.

It is probable that the pattern of feeding in many other monopisthocotyleans does not significantly differ from that described above; but few have been studied in detail. Those species which are known or believed to feed only on epidermis or mucus are listed in table 30. It can be speculated that epidermal feeders could occasionally penetrate deeper and become blood feeders; so it is not surprising to find that haemoglobin has been detected in the gut of a few species e.g. *Amphibdella torpedinis*, *A. flavolineata*, *Dactylogyrus vastator*, *D. solidus*, *Ancylodiscoides parasiluri* and *Tetraonchus* sp.[156, 329]

Polyopisthocotylea. In this group the mouth is terminal, and either is provided with an oral sucker or leads into a buccal cavity with buccal suckers. The buccal cavity leads into a muscular

TABLE 30

Nature of food materials of some Monogenea

	Host	Location	Food	Reference
MONOPISTHOCOTYLEA				
Entobdella hipoglossi	halibut	skin	epidermis	128a
E. soleae	sole	skin	epidermis	157
Calicotyle kröyeri	skate	cloaca	epithelium, mucus	128a
Udonella caligorum	copepod	skin	mucus	128a
Leptocotyle minor	dogfish	skin	*epidermis	187
Calceostoma calceostoma	*Argyrosomus regium*	skin	*epidermis	157
Gyrodactylus spp.	goldfish	skin, gills	*epithelium	157
Diplectanum aequans	bass	gills	*epithelium	157
Amphibdelloides maccallumi	*Torpedo nobiliana*	gills	*epithelium	157
Amphibdella torpedinis	*T. marmorata*	gills	†blood	157
A. flavolineata	*T. nobiliana*	gills	†blood	157
Capsala martinieri	sunfish	skin	*epidermis	157
Trochopus sp.	*Trigla* sp.	gills	*epidermis	157
Nitzschia sturionis	sturgeon	gills	‡epidermis? blood?	157
Acanthocotyle spp.	*Raia* spp.	skin	epidermis	157
Dactylogyrus vastator	carp?	gills	tissue, mucus, blood	329
Ancylodiscoides parasiluri	?	gills	tissue, mucus, blood	329
POLYOPISTHOCOTYLEA				
Polystoma integerrimum	frog	ur. bladder	blood	148
Hexabothrium appendiculata	dogfish	gills	blood	187
Octostoma scombri	mackerel	gills	blood	187
Discocotyle sagittata	trout	gills	blood	128a
Anthocotyle merluccii	hake	gills	blood	187
Axine belones	gar-fish	gills	blood	187
Diclidophora merlangi	whiting	gills	blood, tissue, mucus	128a
Plectanocotyle gurnardi	gurnard	gills	blood	128a
Diplozoon paradoxum	minnow	gills	blood, tissue	128a
Octodactylus palmata	ling	gills	blood, tissue, mucus	128a

* Based on indirect evidence such as absence of blood pigment. (Kearn[157])
† Other tissue probably ingested also.
‡ Evidence conflicting.

pharynx, the whole being well adapted to the blood-sucking habit.

In these blood-feeding forms, digestion can be either entirely intracellular, as in *Dactylogyrus vastator*,[329] or both extra- and intracellular. In the majority of species, digestion is a combination of intralumenal and intracellular digestion. Ingested blood is very rapidly haemolyzed, and whole and degenerating erythrocytes can be detected in the lumen. After being rendered soluble, food from the lumen enters the gastrodermis by absorption only. The globin moiety forms the chief nutriment for the parasite and the haematin is not used for nutritional purposes. This pigment appears unaltered in the gut cells and is eliminated either by direct discharge from them into the lumen, or by the sloughing off of intact epithelial cells.[148, 187] Haematin can be detected in the urine of frogs infected with *Polystoma integerrimum* and its presence has served to identify infected frogs.[147] The method is to squeeze a little urine into a tube to which is added an alkaline solution of luminal and hydrogen peroxide. The presence of haematin, and hence *Polystoma*, is indicated by the development of an intense blue luminescence.

Digestive enzymes. A protease, lipase, aminopeptidase, esterase and an alkaline phosphatase have been reported to be present in some species of Monogenea.[127]

Metabolism

General. Since many monogeneans are external parasites of marine fish, it is likely that they could be maintained for considerable periods in conditions approaching those *in vivo*. This would enable some metabolic determinations to be carried out. Yet few attempts to culture monogeneans have been recorded, and metabolic studies on respiration, carbohydrate, protein or fat metabolisms do not appear to have been made. Examination of egg-production in some 13 species of Monogenea showed that at 3–7° C. adult worms survived for 2–3 weeks but little egg-production took place below 8° C.[190] At 13° C., which is approximately the *in vivo* temperature, egg-production generally ceased after 4 days, suggesting that the condition of the worms rapidly degenerates at this temperature. At 18° C. parasites did not survive longer than 24 hours but egg-production continued for the first 12 hours.

At 20° C. parasites died within 12 hours, generally without egg capsules being produced.

Carbohydrate metabolism. Since monogeneans are primarily ecto-parasites and have access to oxygen, it can be expected that their metabolism will be primarily aerobic and their food reserves lower than those of the digeneans, which are primarily endo-parasitic, anaerobic, organisms. The few studies carried out[127, 219]

TABLE 31

Chemical composition of some Monogenea (after Halton[127])

Species	Host	Location	dry wt. as % F.W.	inorganic ash	glycogen	protein	iron
				Chemical composition as % dry wt.			
MONOPISTHOCOTYLEA							
Entobdella hippoglossi	halibut	skin	16	5·7	0·8	83	.
Calicotyle kröyeri	skate	cloaca	12	4·9	1·17	79	.
POLYOPISTHOCOTYLEA							
Polystoma integerrimum	frog	ur. bladder	9·5	7·2	8·5	67	0·8
Discocotyle sagittata	trout	gills	9·5	6·0	0·9	.	0·4
Diclidophora merlangi	whiting	gills	7·5	7·0	1·2	74	0·65
Octodactylus palmata	ling	gills	7	6·5	2·5	79	0·7

in general bear this out and are summarized in table 31. In endo-parasitic monogeneans, however, the glycogen reserves approach those of Digenea. Glycogen reserves are stored in the parenchyma as in Digenea. Related undoubtedly to their blood feeding habits, substantial quantities of iron are found in polyopisthocotyleans.

Respiration. No figures for oxygen consumption of mono-geneans are available. The majority of species, being parasitic on the gills, will have ready access to abundant oxygen—indeed, probably more than is normally available even to the majority of free-living turbellarians. Some species, however, live in habitats

deficient in oxygen and at least one species appears to have developed special mechanisms for obtaining oxygen.[155] This is *Entobdella
soleae*, which is found on the blind (='ventral') surface of the
common sole (*Solea solea*), a fish which spends most of the day
buried (apart from the eye-bearing head surface) in the sea bottom,
and swims freely only at night.[156] If there are no disturbing

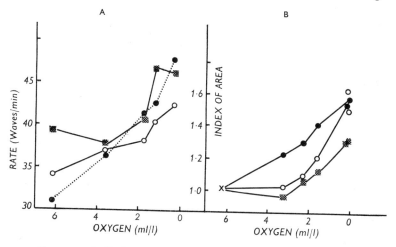

Fig. 59.   A. Relationship between rate of body undulation
and O_2 content of medium in 3 individuals of *Entobdella
soleae*.   B. Relationship between 'surface area' and O_2
content of medium in 3 other individuals of *E. soleae*. (after
Kearn, 1962)

currents, the ocean bottom is generally deficient in oxygen, so that
the parasite is in a medium of low oxygen tension. In substrates of
low oxygen tension, *E. soleae* develops characteristic undulating
movements which consist of waves transverse to the body passing
along the parasite from posterior to anterior. The parasite
attaches with the haptor nearer to the anterior end of the fish. In
normally oxygenated sea water (containing about 6 ml. O_2/l.), the
undulations were recorded at 35 waves per min. at 10° C. The
Q_{10} for the undulation rate was approximately 2.

It may be demonstrated (fig. 59) that, if the parasite is placed
in sea water of decreasing oxygen content, the undulation rate is
inversely proportional to the oxygen content. The parasite

however, adjusts itself to changing oxygen conditions not only by varying its rate of undulation but also by changing the amplitude of the wave and the exposed surface area of its body. There is no evidence to indicate how these adaptations are coordinated. These breathing movements can perhaps be compared to 'tail' movements of the mud-dwelling tubificid worms. Undulating movements similar to those of *Entobdella* have been described in three other skin-dwelling monogeneans also, *Acanthocotyle* sp., *Pseudocotyle squatinae* and *Leptocotyle minor*.

Nervous system

The neurophysiology of the Monogenea has not been studied. The presence of cholinesterase in the nervous system of *Diplozoon paradoxum* has been strikingly demonstrated by histochemical means (plate IV).[128] In contrast, the closely related species *Discocotyle sagittata* gives a negative reaction for esterases with this technique. The basis for these differences is not understood.

Reproduction

Maturation. Although there have been numerous detailed anatomical studies on the development of monogeneans, very little is known about the physiology of reproduction or about the metabolism or the physiological basis of the behaviour of the larval forms. Since parasites often depend on larval hosts for future infections, it is not surprising to find that host and monogenean life cycles are closely synchronized. The classical example of this is *Polystoma integerrimum*, whose genitalia mature only as its frog host is preparing to enter the water preparatory to copulation. When frogs enter the water and copulation is proceeding, maturation of *Polystoma* takes place and numerous eggs are produced. It can be speculated that the maturation processes in the fluke are directly or indirectly under the control of the hormonal activity of the frog host. Thus, when infected immature frogs were injected with pituitary extract, the monogeneans in the bladder were found to mature within 4–8 days and produce eggs for about a week. After this period—which corresponds approximately to the time frogs spend on spawning—no further eggs were produced.[221]

PLATE IV. *Diplozoon paradoxum*, showing localization of cholinesterase in the nervous system; indoxyl acetate technique. (after Halton and Jennings, 1964)

The nature of the endocrinal control has not been established. Clearly, it could be a result of increased levels of either gonado-tropins or gonadal hormones; for both these would occur after pituitary injection. Since *Polystoma* feeds on blood, it would have ready access to these hormones. It is interesting to note that the protozoan parasites *Opalina* and *Nyctotherus* in the rectum of the same host have a life cycle similarly synchronized with that of the host. In the case of *Opalina*, the experimental evidence suggests that it is the gonadal, rather than the pituitary, hormones which are the controlling factors.[223] Maturation of *Polystoma* may also be brought about by keeping frogs at 0–1° C. for 5–6 days and then keeping at 20–25° C. for 1–5 days.[164] This result may, of course, be related to a change in hormonal levels of the host brought about by the temperature shock. In frogs kept at 20–25° C. without a previous cold shock, maturation did not occur! In other amphibian species also, with a life cycle comparable with that of *Rana temporaria*, it is likely that a synchronized cycle will occur. There is some suggestive experimental evidence that *Polystoma stellai* in the frog *Hyla septentrionalis* follows this pattern.[312] In contrast, polystomes which are parasitic in species of amphibia which remain in water all the year round (e.g. *Protopolystoma xenopi* in *Xenopus laevis*) mature normally and eggs are shed all the year round.[320]

It is interesting to note also that the infection of *P. integerrimum* is higher in male frogs than in female frogs just before the breeding season. This suggests that, in some way not understood, a high level of female gonadal hormones reduces the level of parasitiza-tion by *Polystoma*. The effect is noted also with other helminth parasites in these hosts, so that a general effect against parasites as such, rather than against monogeneans in particular, is apparently involved.[176, 177]

A cycle of sexual activity has also been reported in *Gastrocotyle trachuri* on the gills of *Trachurus trachurus*, the parasites becoming reproductively inactive just before the departure of the bottom-feeding young scad in July (to become plankton feeders), and becoming active again in October just before the descent to the bottom of young newly hatched scad. The physiological mechan-ism regulating this cycle is unknown, but it has been conjectured that a state of diapause may occur.[193]

Fertilization. Although the mechanics of fertilization and copulation have been studied in the Monogenea, the physiology and biochemistry of these processes have not been examined. Cross- or self-fertilization can occur and in some cases spermatophores are employed. In *Entobdella diadema* these are produced by a pair of glands lying one on each side of the body beside the main nerve cord; the chemical nature of the spermatophore is not known.[196] Detailed accounts of copulation in several species are given by Bychowsky.[29] In *P. integerrimum* the behaviour pattern is unusual in that about twenty copulations can occur within an hour, the copulating worms alternately playing the rôle of male and female respectively. In this species, the vaginal apertures on each side of the body are in the form of sieve plates, and only the sclerotized hooks of the cirrus can be introduced. That self-fertilization can occur is proved by the fact that single individuals of *Polystoma* in a bladder can produce fertile eggs.

The most remarkable adaptation to cross-fertilization is probably found in the well-known *Diplozoon paradoxum* (plate IV), which consists of two individuals fused together. Attachment takes place in the larval stage, a pair being held together by a sucker arrangement which finally disappears as the organisms become fused together. In mature individuals, the vagina of one individual opens opposite the openings of the uterus and vas deferens of the other, so that cross-fertilization can occur. Individual larvae which fail to find a partner subsequently die. The fusion of larval individuals in this way raises interesting biochemical, genetical and immunological questions. It would be interesting to know, for example, whether only genetically identical individuals can fuse or whether ' strains ' from different hosts or from different ecological areas can do so.

Egg-shell formation. The process of egg-shell formation[29, 61] does not appear to differ significantly, in its broad outline, from that in the Digenea.[292] The bulk of the shell capsule material comes from the vitelline cells, with perhaps a minor contribution from Mehlis's gland, whose cells are of two types.[164] In the few species examined—*Protopolystoma xenopi*;[320] *Kuhnia* sp.;[150] *P. integerrimum*;[164]—the capsule is a quinone tanned protein (see p. 61), and the vitellaria give positive reactions for phenolase and polyphenols. Both these substances may be demonstrated histochemically in whole mounts, the latter in particular giving striking microscopical preparations with the Fast Red Salt B technique for polyphenols[150] (plate I).

Egg production. Monogeneans are capable of producing large numbers of eggs but the number varies considerably. *P. integerrimum* deposits about 2000–2500 during its short laying season. Egg-production appears to be increased by a rise in temperature and by a fall in oxygen levels; the reason for the latter effect is not understood. Although some eggs are simple, like those in the majority of Digenea, many have long anterior and posterior filaments.

Embryonation and hatching. The pattern of embryonation appears not to differ significantly from that in digeneans. The rate of development increases sharply with temperature (fig. 60) as in other platyhelminths. Embryonation in species so far studied is inhibited at 4° C.,[29] but in some species development virtually ceases at 8° C.;[190] this may be an adaptation to water temperatures in different areas. Presumably oxygen is required for embryonation, but there is no information on the level of requirements.

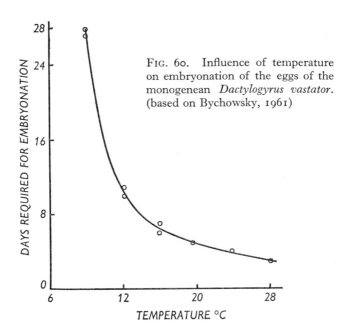

FIG. 60. Influence of temperature on embryonation of the eggs of the monogenean *Dactylogyrus vastator*. (based on Bychowsky, 1961)

The physiology of hatching has been little studied. Light may initiate the hatching process of eggs of some species[29] but whether a hatching enzyme is released, as in the Digenea (p. 70), is not known. If the embryonated eggs of *Squalonchocotyle torpedinis* are left in the water in which they were embryonated, little hatching occurs; but if they are placed in water containing either the host (*Torpedo marmorata*) or even its excised gills, almost complete hatching takes place.[92] Although this experiment clearly needs repeating with adequate controls, it does suggest that the host tissue may release some ' hatching factor ' which triggers off the hatching process in this species. In *Polystomoidella oblonga*, a parasite of the common musk turtle *Stenotherus odoratus*, hatching occurs *in utero* (ovoviviparity) and the released larvae, which are non-ciliated and in an advanced state of development, probably develop directly into adults without a free-living stage.[234] This reproductive behaviour could account for the fact that large numbers of individuals of this species (up to thirty) are sometimes found in the host.

Larval behaviour. The hatched larva of a monogenean is an *oncomiracidium*, the characteristic features of which are shown in fig. 61. Almost nothing is known of the metabolism of these larvae, and very little about their behaviour. The oncomiracidia of most species are heavily ciliated and bear numerous posterior hooks, so that they are well adapted both for swimming and for attachment. Well-developed head glands are present. Many species show positive phototropism, presumably related to the presence of eye spots, which are well developed in many cases, although some (e.g. *Diclidophora denticulata*), lack such structures and show no reaction to light.[95] The length of larval life is about twenty-four hours; hence glycogen reserves cannot be large. Larvae which cease swimming are reported to crawl on the bottom for some time.

Oncomiracidia appear to find their hosts by chance and the well-developed posterior hooks mean that they are well adapted for attachment. Little is known of the physiological processes of invasion of hosts or growth of larvae. The factors influencing growth are not known, but they may prove to be somewhat unusual with some species. This is borne out by the well-known example of the larvae of *P. integerrimum*, which mature in different times

and in somewhat different ways, depending on the site of attachment. Larvae which become attached to the external gills mature
within about twenty days[61] and produce eggs which give rise to
miniature worms. These differ from the normal adult (matured in
the bladder) in having only a single testis, functionless copulatory

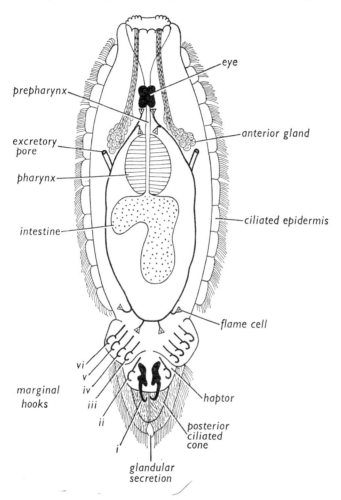

FIG. 61. Oncomiracidium of a monogenean, *Gastrocotyle
trachuri* from the gills of *Trachurus trachurus*. (after
Llewellyn, 1963)

organs, headless sperms and rudimentary uterus, vaginae and vitellariae. Clearly, morphogenesis in such forms is ' abnormal ', although the eggs produced give rise to normal larvae. Larvae which attach to the internal gills pass down the alimentary canal when the tadpole metamorphoses, and so reach the bladder. In this site maturation is very slow, taking up to three years. The physiological basis of the differences between these two types of development is not understood.

Monogeneans as experimental material

From the above brief survey it is clear that knowledge of the physiology of the Monogenea is extremely limited. Yet, like the digenetic trematodes, monogeneans would appear to offer excellent material for experimental studies. They have the advantage over digeneans that they do not require an intermediate host and, being mainly external parasites, are also likely to require less rigid environmental conditions for *in vitro* culture. Biochemical problems concerning this group abound, since the metabolic pathways are entirely unknown. In addition, problems of exceptional biological interest occur, such as the apparent hormonal relationship between *Polystoma* and its frog host, and the genetical and immunological problems which arise when two larvae of *Diplozoon* fuse to form one individual. The behaviour and possible tropisms of the oncomiracidia also offer scope for further work, as does the physiology of egg development and hatching. Many of these problems are common to other groups of platyhelminths, and their solution could contribute substantially to our general knowledge of fundamental biological processes.

Appendix

TABLE 32

Principal species of Digenea referred to in this volume

(Intermediate and definitive hosts, other than those listed, may be utilized in different countries.)

Species	Definitive host(s)	Normal location	First intermediate host	Second intermediate host
Alaria arisaemoides	carnivores	intestine	*Planorbula armigera*	tadpoles*
Alaria canis	carnivores	intestine	*Helisoma trivolvis*	tadpoles*
Austrobilharzia terrigalensis	birds	int. veins	*Velacumantus (Pyrazus) australis*	—
Bucephalopsis gracilescens	angler fish	gut	*Cardium edule*	fish
Clonorchis sinensis	carnivores, man	bile ducts	*Parafossarulus* spp.	fish
Coitocaecum anaspidis	fish?	unknown	unknown	Tasmanian shrimp†
Cryptocotyle lingua	birds, mammals	intestine	*Littorina littorea*	fish
Cyathocotyle bushiensis	ducks	caeca	*Bithynia tentaculata*	*Bithynia tentaculata*
Dicrocoelium dendriticum	sheep	bile ducts	*Cionella lubrica*	ants
Diplostomum flexicaudum	birds	intestine	*Lymnaea* spp.	fish
Diplostomum phoxini	ducks	intestine	*Lymnaea pereger*	minnow
Fasciola gigantica	ruminants	bile duct	*Lymnaea natalensis*	encysts on vegetation
Fasciola hepatica	sheep	bile ducts	*Lymnaea truncatula*	encysts on vegetation
Fascioloides magna	deer, cattle	liver	*Fossaria* spp.	encysts on vegetation

TABLE 32—*continued*

Species	Definitive host(s)	Normal location	First intermediate host	Second intermediate host
Fasciolopsis buski	man, pigs	intestine	*Planorbidae*	encysts on vegetation
Gastrothylax crumenifer	buffalo	rumen	?	encysts on vegetation
Gigantobilharzia gyrauli	birds	int. veins	*Gyraulus parvus*	—
G. huronensis	birds	int. veins	*Physa gyrina*	—
G. huttoni	birds	int. veins	*Haminoea antillarum*	—
Glypthelmins quieta	amphibia	intestine	*Physa gyrina*	frogs
Gorgodera amplicava	amphibia	bladder	*Musculium partumeium*	*Helisoma* spp. tadpoles, etc.
Gorgoderina vitelliloba	amphibia	bladder	*Sphaerium* spp.	tadpoles, alder flies
Gynaecotyla adunca	birds (also fish and turtles)	small intestine	*Nassa vibex*	fiddler crab
Haematoloechus variegatus	frogs	lungs	*Lymnaea* spp.	dragonflies
Halipegus occidualis	frogs	buccal cavity	*Helisoma antrosa*	*Cyclops* spp. (3rd int. host dragonflies)
Haplometra cylindracea	frogs	lungs	*Lymnaea* spp.	*Ilybius fuliginosus*
Heterobilharzia americana	mammals	mes. veins	*Pseudosuccinea columella*	—
Himasthla quissetensis	herring gull	intestine	*Nassa obsoleta*	lamellibranchs
Notocotylus ralli	water rail	caecum	*Tropodiscus carinatus*	encysts on vegetation
N. urbanensis	birds, mammals	caecum	*Stagnicola emarginata*	encysts on vegetation
Opisthioglyphe ranae	amphibia	intestine	*Lymnaea* spp.	tadpoles
Paragonimus ohirai	mammals	lungs	*Assiminea* spp.	crabs
P. westermani	man, carnivores, rodents	lungs	*Semisulcospira libertina*	crustaceans
Paramphistomum cervi	cattle	rumen	*Bulinus* spp.	encysts on vegetation

TABLE 32—*continued*

Species	Definitive host(s)	Normal location	First intermediate host	Second intermediate host
Parorchis acanthus	birds	rectum, bursa Fabricii	*Purpura lapillus*	encysts on surfaces
Philophthalmus gralli	birds	eye	*Tarebia granifera*	encysts in snail
Posthodiplostomum cuticola	birds	intestine	*Tropodiscus* spp.	fish
Proctoeces subtenuis	fish	hind gut	lamellibranchs†	*Scrobicularia plana*
Prosthodendrium chilostomum	bats	intestine	*Lymnaea* spp.	caddis fly larvae
Saccocoelium beauforti	mullet	intestine	?	?
Schistosoma haemotobium	man, other mammals	mes. veins	*Oncomelania* spp.	—
S. japonicum	man, other mammals	mes. veins	*Bulinus* spp.	—
S. mansoni	man, other mammals	mes. veins	*Biomphalaria* spp.	—
Schistosomatium douthitti	rodents	portal veins	*Lymnaea* spp.	—
Trichobilharzia ocellata	birds	submuco-sal venous plexus	*Lymnaea* spp.	—
Uvulifer ambloplitis		intestine	*Helisoma trivolvis*	fish
Zoogonus rubellus	fish	intestine	*Nassarius obsoleta*	—

* Developmental stages are *mesocercariae* (essentially prolonged cercarial stages); metacercariae (diplostomula) develop in the lungs of carnivores.

† Metacercariae progenetic.

References

1. ABDEL-MALEK, E. T. 1950. Susceptibility of the snail *Biomphalaria boissyi* to infection with certain strains of *Schistosoma mansoni*. *Am. J. trop. Med.*, **30**: 887–894.
2. ALICATA, J. E. 1962. Life cycle and developmental stages of *Philophthalmus gralli* in the intermediate and final hosts. *J. Parasit.*, **48**: 47–54.
3. ANDERSON, M. G. 1962. *Proterometra dickermani sp. nov.* (Trematoda: Azygiidae.) *Trans. Am. microsc. Soc.*, **81**: 279–282.
4. ANDRADE, Z. A. and BARKA, T. 1962. Histochemical observations on experimental schistosomiasis of mouse. *Am. J. trop. Med. Hyg.*, **11**: 12–16.
5. ARVY, L. 1954. Distomatose cérébro-rachidienne due à *Diplostomulum phoxini* (Faust), Hughes, 1929, chez *Phoxinus laevis* A G. *Annls Parasit. hum. comp.*, **29**: 510–520.
6. AXMANN, M. C. 1947. Morphological studies on glycogen deposition in schistosomes and other flukes. *J. Morph.*, **80**: 321–344.
7. BAER, J. and JOYEUX, C. 1961. ' Plathelminthes, Mésozoaires, Acanthocéphales, Némertiens.' In: *Traité de Zoologie*. (*Ed.* P. Grassé.) Vol. IV. Fasc. 1 S. 561–692. Masson et Cie., Paris.
8. BAILLY-CHANTEMERGUE, S. 1957. Structure cuticulaire de *Diplostomulum phoxini* (Faust) (Trematoda Diplostomidae). *Archs Zool. exp. gén.*, **94**: 17–42.
9. BELL, E. J. and SMYTH, J. D. 1958. Cytological and histochemical criteria for evaluating development of trematodes and pseudophyllidean cestodes *in vivo* and *in vitro*. *Parasitology*, **48**: 131–148.
10. BERRIE, A. D. 1960. The influence of various definitive hosts on the development of *Diplostomum phoxini* (Strigeida, Trematoda). *J. Helminth.*, **34**: 205–210.
11. BIGUET, J., CAPRON, A. and TRAN VAN KY, P. 1962. Les antigènes de *Fasciola hepatica*. Étude électrophorétique et immunoélectrophorétique. Identification des fractions et comparaison avec les antigènes correspondant à sept autres helminthes. *Annls Parasit. hum. comp.*, **37**: 221–231.
12. BISSET, K. A. 1947. Bacterial infection and immunity in lower vertebrates and invertebrates. *J. Hyg., Camb.*, **45**: 128–135.
13. BJÖRKMAN, N. and THORSELL, W. 1964. On the fine structure and resorptive function of the cuticle of the liver fluke, *Fasciola hepatica*, L. *Expl Cell Res.*, **33**: 319–329.
14. BOGITSH, B. J. 1962. The chemical nature of metacercarial cysts. I. Histological and histochemical observations on the cyst of *Posthodiplostomum minimum*. *J. Parasit.*, **48**: 55–60.
15. —1963. Histochemical observations on the cercariae of *Posthodiplostomum minimum*. *Expl. Parasit.*, **14**: 193–202.
16. BOGOMOPOVA, N. A. and CHAVPOVA, P. E. 1961. [Vitelline cells of *Fasciola hepatica* and *Diphyllobothrium latum* and their role in the formation of the egg-shell and nutrition of the embryo.] *Helminthologia*, **III**: 47–59.

17. BROWN, F. J. 1926. Some British freshwater larval trematodes, with contributions to their life histories. *Parasitology*, **18**: 21–34.
18. BRÜEL, D., HOLTER, H., LINDERSTRØM-LANG, K. and ROZITS, K. 1946. A micromethod for the determination of total nitrogen (accuracy 0·005 μg N). *C. r. Trav. Lab. Carlsberg, Sér. chim.*, **25**: 289–324.
19. BRUIJNING, C. F. A. 1961. The circumoval precipitin test in experimental *Schistosoma mansoni* infections. *Trop. geogr. Med.*, **13**: 378–387.
20. BRUSKIN, B. R. 1959. [A histochemical investigation of glycogen in *Opisthorchis felineus.*] *Dokl. Akad. Nauk SSSR*, **127**: 1315–1316. (In Russian.)
21. BRYANT, C. and WILLIAMS, J. P. G. 1962. Some aspects of the metabolism of the liver fluke, *Fasciola hepatica*, L. *Expl Parasit.*, **12**: 372–376.
22. BUEDING, E. 1949. Metabolism of parasitic helminths. *Physiol. Rev.*, **29**: 195–218.
23. —1950. Carbohydrate metabolism of *Schistosoma mansoni*. *J. gen. Physiol.*, **33**: 475–495.
24. —1952. Acetylcholinesterase activity of *Schistosoma mansoni*. *Br. J. Pharmac. Chemother.*, **7**: 563–566.
25. —1962. Comparative biochemistry of parasitic helminths. *Comp. Biochem. Physiol.*, **4**: 343–357.
26. BUEDING, E. and CHARMS, B. 1951. Respiratory metabolism of parasitic helminths without participation of the cytochrome system. *Nature, Lond.*, **167**: 149.
27. BURNET, F. M. 1959. *The Clonal Selection Theory of Acquired Immunity.* Cambridge University Press.
28. BURTON, P. R. 1964. The ultrastructure of the integument of the frog lung-fluke, *Haematoloechus medioplexus* (Trematoda: Plagiorchiidae). *J. Morph.*, **115**: 305–318.
29. BYCHOWSKY, B. E. 1957. *Monogenetic Trematodes, their Systematics and Phylogeny.* (English ed. 1961) Am. Inst. Biol. Sci., Washington.
30. CALLOT, J. 1939. Particularités biologiques de la metacercaire de *Posthodiplostomum cuticola* (von Nordmann). *Annls Parasit. hum. comp.*, **17**: 332–335.
31. CAMPBELL, W. C. 1960. Presence of phenolase in the fasciolid metacercarial cyst. *J. Parasit.*, **46**: 848.
32. 1961. Notes on the egg and miracidium of *Fascioloides magna* (Trematoda). *Trans. Am. microsc. Soc.*, **80**: 308–319.
33. CAMPBELL, W. C. and TODD, A. C. 1955. Behaviour of the miracidium of *Fascioloides magna* (Bassi, 1875) Ward, 1917, in the presence of a snail host. *Trans. Am. microsc. Soc.*, **74**: 342–347.
34. CARDELL, R. R. 1962. Observations on the ultrastructure of the body of the cercaria of *Himasthla quissetensis* (Miller and Northup, 1926). *Trans. Am. microsc. Soc.*, **81**: 124–131.
35. CHEEVER, A. W. and WELLER, T. H. 1958. Observations on the growth and nutritional requirements of *Schistosoma mansoni in vitro*. *Am. J. Hyg.*, **68**: 322–339.
36. CHENG, T. C. 1962. The effects of *Echinoparyphium* larvae on the structure of and glycogen deposition in the hepatopancreas of *Helisoma trivolvis* and glycogenesis in the parasite larvae. *Malacologia* **1**, 291–303.

37. —1963*a*. Biochemical requirements of larval trematodes. *Ann. N.Y. Acad. Sci.*, **113**: 289–320.

38. —1963*b*. Histological and histochemical studies on the effects of parasitism of *Musculium partumeium* (Say) by the larvae of *Gorgodera amplicava* Looss. *Proc. helminth. Soc. Wash.*, **30**: 101–107.

39. —1963*c*. Activation of *Gorgodera amplicava* cercariae by molluscan sera. *Expl Parasit.*, **13**: 342–347.

40. —1963*d*. One aspect of the biochemical basis for development: the source and utilization of amino acids in intramolluscan larval trematodes. In: *Factors Governing Morphogenesis of Parasitic Animals.* (Ed. T. C. Cheng.) Pa Acad. Sci., Pittsburg.

41. CHENG, T. C. and SANDERS, B. G. 1962. Internal defense mechanisms in molluscs and an electrophoretic analysis of a naturally occurring serum hemagglutinin in *Viviparus malleatus* Reeve. *Proc. Pa Acad. Sci.*, **36**: 72–83.

42. CHENG, T. C. and SNYDER, R. W., Jr. 1962*a*. Studies on host-parasite relationships between larval trematodes and their hosts. I. A review. II. The utilization of the host's glycogen by the intra-molluscan larvae of *Glypthelmins pennsylvaniensis* Cheng, and associated phenomena. *Trans. Am. microsc. Soc.*, **81**: 209–228.

43. —1962*b*. Studies on host-parasite relationships between larval trematodes and their hosts. III. Certain aspects of lipid metabolism in *Helisoma trivolvis* (Say) infected with the larvae of *Glypthelmins pennsylvaniensis* Cheng and related phenomena. *Trans. Am. microsc. Soc.*, **81**: 327–331.

44. —1963. Studies on host-parasite relationships between larval trematodes and their hosts. IV. A histochemical determination of glucose and its role in the metabolism of molluscan host and parasite. *Trans. Am. microsc. Soc.*, **82**: 343–346.

45. CHENG, T. C. and STREISFELD, S. D. 1963. Innate phagocytosis in the trematodes *Megalodiscus temperatus* and *Haematoloechus sp. J. Morph.*, **113**: 375–380.

46. CHERNIN, E. and DUNAVAN, C. A. 1962. The influence of host-parasite dispersion upon the capacity of *Schistosoma mansoni* miracidia to infect *Australorbis glabratus*. *Am. J. trop. Med. Hyg.*, **11**: 455–471.

47. CH'IEN, L. and BANG, F. B. 1950. Agglutination of cercariae of *Schistosoma mansoni* by immune sera. *Proc. Soc. exp. Biol. Med.*, **74**: 68–72.

48. CHING, H. L. 1961. The development and morphological variation of *Philophthalmus gralli* Mathis & Leger, 1910 with a comparison of species of *Philophthalmus* Looss, 1899. *Proc. helminth. Soc. Wash.*, **28**: 130–138.

49. CHRISTENSEN, H. N. 1960. Reactive sites and biological transport. *Adv. Protein Chem.*, **15**: 239–314.

50. CLEGG, J. A. 1957. Studies on the maintenance of *Fasciola hepatica in vitro*. Ph.D. Thesis, Univ. of London.

51. —1959. Development of sperm by *Schistosoma mansoni* cultured *in vitro*. *Bull. Res. Coun. Israel*, **8E**: 1–6.

52. —1961. A continuous-flow apparatus for *in vitro* culture of *Schistosoma mansoni*. *Bull. Res. Coun. Israel*, **9E**: 3–4.

53. —1965*a*. Secretion of lipoprotein by Mehlis' gland in *Fasciola hepatica*. N.Y. Acad. Sci. (In press.)

54. —1965*b*. *In vitro* cultivation of *Schistosoma mansoni*. *Expl Parasit.*, **16**: 133–147.

55. CLEGG, J. A. and SMYTH, J. D. 1966. Growth, development and culture methods: parasitic platyhelminths. In: *Chemical Zoology* Vol. I, Chapter 1. Academic Press, London and New York.

56. COIL, W. H. and KUNTZ, R. E. 1963. Observations on the histochemistry of *Syncoelium spathulatum n.sp. Proc. helminth. Soc. Wash.*, **30**: 60–65.

57. CROWLE, A. J. 1961. *Immunodiffusion.* Academic Press, London and New York.

58. CRUICKSHANK, R. (*ed.*). 1963. *Modern Trends in Immunology.* Butterworths, London.

59. DAUGHERTY, J. W. 1952. Intermediary protein metabolism in helminths. I. Transaminase reactions in *Fasciola hepatica. Expl Parasit.*, **1**: 331–338.

60. DAVENPORT, D., WRIGHT, C. A. and CAUSLEY, D. 1962. Technique for the study of the behavior of motile microorganisms. *Science, N.Y.*, **135**: 1059–1060.

61. DAWES, B. 1946. *The Trematoda.* Cambridge University Press.

62. 1959. Penetration of the liver-fluke, *Fasciola hepatica*, into the snail, *Limnaea truncatula. Nature, Lond.*, **184**: 1334–1335.

63. —1960*a*. Penetration of *Fasciola gigantica* Cobbold, 1856, into snail hosts. *Nature, Lond.*, **185**: 51–53.

64. —1960*b*. A study of the miracidium of *Fasciola hepatica* and an account of the mode of penetration of the sporocyst into *Limnaea truncatula. Sobretiro del Libro Homenaje al Dr. Eduardo Caballero y Caballero.*

65. —1960*c*. The penetration of *Fasciola hepatica* into *Limnaea truncatula*, and of *F. gigantica* into *L. auricularia. Trans. R. Soc. trop. Med. Hyg.*, **54**: 9–10.

66. —1961*a*. Juvenile stages of *Fasciola hepatica* in the liver of the mouse. *Nature, Lond.*, **190**: 646–647.

67. —1961*b*. The juvenile stages of *Fasciola hepatica* during early penetration into the liver of the mouse. *Trans. R. Soc. trop. Med. Hyg.*, **55**: 310–311.

68. —1961*c*. On the early stages of *Fasciola hepatica* penetrating into the liver of an experimental host, the mouse: a histological picture. *J. Helminth., R. T. Leiper Supplement*, 41–52.

69. —1962. A histological study of the caecal epithelium of *Fasciola hepatica* L. *Parasitology*, **52**: 483 493.

70. —1963*a*. Some observations of *Fasciola hepatica* L. during feeding operations in the hepatic parenchyma of the mouse, with notes on the nature of liver damage in this host. *Parasitology*, **53**: 135–143.

71. —1963*b*. The migration of juvenile forms of *Fasciola hepatica* L. through the wall of the intestines in the mouse, with some observations on food and feeding. *Parasitology*, **53**: 109–122.

72. DAWES, B. and HUGHES, D. L. 1964. Fascioliasis: the invasive stages of *Fasciola hepatica* in mammalian hosts. In: *Advances in Parasitology* (ed. B. Dawes), **2**: 97–168. Academic Press, London.

73. DAWES, B. and MÜLLER, R. 1957. Maintenance *in vitro* of *Haplometra cylindracea. Nature, Lond.*, **180**: 1217.

74. DE MEILLON, B. and PATERSON, S. 1958. Experimental bilharziasis in animals. VII. Effect of a low-protein diet on bilharziasis in white mice. *S. Afr. Med. J.*, **32**: 1086–1088.
75. DINNIK, J. A. and DINNIK, N. N. 1962. The growth of *Paramphistomum microbothrium* Fischoeder to maturity and its longevity in cattle. *Bull. epizoot. Dis. Afr.*, **10**: 27–31.
76. DIXON, K. E. 1964. Excystment of metacercariae of *Fasciola hepatica* L. in vitro. *Nature, Lond.*, **202**: 1240–1241.
77. —1965a. The structure and histochemistry of the cyst wall of the metacercaria of *Fasciola hepatica* L. *Parasitology*, **55**: 215–226.
78. —1965b. Unpublished work.
79. DIXON, K. E. and MERCER, E. H. 1964. The fine structure of the cyst wall of the metacercaria of *Fasciola hepatica*. *Q. Jl. microsc. Sci.*, **105**: 385–389.
80. DIXON, M. and WEBB, E. C. 1964. *Enzymes*. 2nd ed. Longmans, Green, London.
81. DÖNGES, J. 1962. Entwicklungsgeschichtliche und morphologische Untersuchungen an Notocotyliden (Trematoda). *Z. ParasitKde.*, **22**: 43–67.
82. —1963. Reizphysiologische Untersuchungen an der Cercarie von *Posthodiplostomum cuticola* (v. Nordmann 1832) Dubois 1936, dem Erreger des Diplostomatiden-Melanoms der Fische. *Verh. dt. zool. Ges.*, 216–223.
83. —1964. Der Lebenszyklus von *Posthodiplostomum cuticola* (v. Nordmann 1832) Dubois 1936 (Trematoda, Diplostomatidae). *Z. ParasitKde.*, **24**: 169–248.
84. DUKE, B. O. L. 1952. On the route of emergence of the cercariae of *Schistosoma mansoni* from *Australorbis glabratus*. *J. Helminth.*, **26**: 133–146.
85. DUSANIC, D. G. 1959. Histochemical observations of alkaline phosphatase in *Schistosoma mansoni*. *J. infect. Dis.*, **105**: 1–8.
86. ERASMUS, D. A. 1959. The migration of cercaria X Baylis (Strigeida) within the fish intermediate host. *Parasitology*, **49**: 173–190.
87. —1964a. Studies on the biology of *Cyathocotyle bushiensis* Khan 1962. *Parasitology*, **16P**.
88. —1964b. Electron microscope and histochemical studies on the cuticle and subcuticular structures of the trematode *Cyathocotyle bushiensis*. *Proc. 1st Int. Congr. Parasit.*, Rome.
89. ERASMUS, D. A. and ÖHMAN, C. 1963. The structure and function of the adhesive organ in strigeid trematodes. *Ann. N.Y. Acad. Sci.*, **113**: 7–35.
90. ERHARDOVÁ, B. 1961. Vývojový cyklus motolice obrovské *Fasciola magna* v podmínkách ČSSR. *Zool. Listy*, **10**: 9–16.
91. ETGES, F. J. and DECKER, C. L. 1963. Chemosensitivity of the miracidium of *Schistosoma mansoni* to *Australorbis glabratus* and other snails. *J. Parasit.*, **49**: 114–116.
92. EUZET, L. and RAIBAUT, A. 1960. Le développement postlarvaire de *Squalonchocotyle torpedinis* (Price 1942) (Monogenea, Hexabothriidae). *Bull. Soc. neuchâtel. Sci. nat.*, **83**: 101–108.
93. FAUST, E. C. 1924. The reactions of the miracidia of *Schistosoma japonicum* and *S. haematobium* in the presence of their intermediate hosts. *J. Parasit.*, **10**: 199–204.

94. FERGUSON, M. S. 1943. Migration and localization of an animal parasite within the host. *J. exp. Zool.*, **93**: 375–401.
95. FRANKLAND, H. M. T. 1955. The life history and bionomics of *Diclidophora denticulata* (Trematoda: Monogenea). *Parasitology*, **45**: 313–351.
96. FRANKLIN, E. C. 1964. The immune globulins—their structure and function and some techniques for their isolation. *Prog. Allergy*, **8**: 58–148.
97. FRANZÉN, A. 1956. On spermiogenesis, morphology of the spermatozoon, and biology of fertilization among invertebrates. *Zool. Bidr. Upps.*, **31**: 356–482.
98. FREEMAN, R. F. H. 1962. Volumetric respirometer measurements of the oxygen consumption of the digenetic trematode *Proctoeces subtenuis*. *Comp. Biochem. Physiol.*, **7**: 199–209.
99. —1963. Haemoglobin in the digenetic trematode *Proctoeces subtenuis* (Linton). *Comp. Biochem. Physiol.*, **10**: 253–256.
100. FRIED, B. 1962. Growth of *Philophthalmus* sp. (Trematoda) on the chorioallantois of the chick. *J. Parasit.*, **48**: 545–550.
101. FRIED, B. and PENNER, L. R. 1963. Exposure of chicks to the metacercaria of *Philophthalmus hegeneri*. *J. Parasit.*, **49**: 978–980.
102. FRIEDL, R. E. 1960. Induced hatching of operculate eggs. *J. Parasit.*, **46**: 454.
103. GEIMAN, Q. M. 1964. Comparative physiology: mutualism, symbiosis, and parasitism. *A. Rev. Physiol.*, **26**: 75–108.
104. GERZELI, G. 1960. Prime osservazioni comparative sulle possibilità di estrazioni selettive dai vitellogeni di *Distoma hepaticum*. *Rc. Ist. lomb. Sci. Lett.*, **B, 94**: 229–246.
105. GINETSINSKAYA, T. A. 1960. [Glycogen in cercariae and the dependence of its distribution on the specific characters of the parasite.] *Dokl. Akad. Nauk SSSR*, **135**: 1012–1015. (In Russian.)
106. —1961. [The dynamics of fat deposition in the life-cycle of trematodes.] *Dokl. Akad. Nauk SSSR*, **139**: 1016–1019. (In Russian.)
107. GINETSINSKAYA, T. A. and DOBROVOLSKI, A. A. 1962. [Glycogen and fat in various phases of the life-cycle of trematodes. I. Morphology of the distribution of glycogen and fat.] *Vest. leningr. gos. Univ. Ser. biol.*, **17**: 67–81. (In Russian.)
108. —1963. [Glycogen and fat in various phases of the life-cycle of trematodes. II. Biological significance of glycogen and fat.] *Vest. leningr. gos. Univ. Ser. biol.*, **18**: 23–33. (In Russian.)
109. GOIL, M. M. 1957. Carbohydrate metabolism in trematode parasites. *Z. ParasitKde.*, **18**: 36–39.
110. —1958a. Fat metabolism in trematode parasites. *Z. ParasitKde.*, **18**: 321–323.
111. —1958b. Protein metabolism in trematode parasites. *J. Helminth.*, **32**: 119–124.
112. —1958c. Rate of oxygen consumption in trematode parasites. *Z. ParasitKde.*, **18**: 435–440.
113. —1959. Haemoglobin in trematodes—*Gastrothylax crumenifer*. *Z. ParasitKde.*, **19**: 362–363.
114. —1961a. Physiological studies on trematodes—*Fasciola gigantica*. Rate of oxygen consumption. *Z. ParasitKde.*, **20**: 568–571.

115. —1961*b*. Physiological studies on trematodes—*Fasciola gigantica*, carbohydrate metabolism. *Parasitology*, **51**: 335–337.

116. —1961*c*. Haemoglobin in trematodes. 1. *Fasciola gigantica*. 2. *Cotylophoron indicum*. *Z. ParasitKde.*, **20**: 572–575.

117. GOLDBERG, E. 1957. Studies on the intermediary metabolism of *Trichinella spiralis*. *Expl Parasit.*, **6**: 367–382.

118. GÖNNERT, R. 1955. Schistosomiasis Studies. II. Über die Eibildung bei *Schistosoma mansoni* und das Schicksal der Eier im Wirtsorganismus. *Z. Tropenmed. Parasit.*, **6**: 33–52.

119. —1962. Histologische untersuchungen über den Feinbau der Eibildungstätte (Oogenotop) von *Fasciola hepatica*. *Z. ParasitKde.*, **21**: 475–492.

120. GOODCHILD, C. G. 1958. Implantation of *Schistosomatium douthitti* into the eyes of rats. *Expl Parasit.*, **7**: 152–164.

121. GORDON, R. M. and GRIFFITHS, R. B. 1951. Observations on the means by which the cercariae of *Schistosoma mansoni* penetrate mammalian skin, together with an account of certain morphological changes observed in the newly penetrated larvae. *Ann. trop. Med. Parasit.*, **45**: 227–243.

122. GRESSON, R. A. R. and PERRY, M. M. 1961. Electron microscope studies of spermateleosis in *Fasciola hepatica* L. *Exp. Cell Res.*, **22**: 1–8.

123. GRESSON, R. A. R. and THREADGOLD, L. T. 1959. A light and electron microscope study of the epithelial cells of the gut of *Fasciola hepatica* L. *J. biophys. biochem. Cytol.*, **6**; 157–162.

124. GRIFFITHS, R. B. 1953. Further observations on the penetration of mammalian skin by the cercariae of *Schistosoma mansoni* with special reference to the effects of mass invasion. *Ann. trop. Med. Parasit.*, **47**: 86–94.

125. HALL, M. C. 1929. Arthropods as intermediate hosts. *Smithson. misc. Collns*, **81**: 1–77.

126. HALTON, D. W. 1963. Some hydrolytic enzymes in two digenetic trematodes. *Proc. XVI Int. Congr. Zool., Wash.*, **1**: 29.

127. —1964. Observations on the nutrition of certain parasitic flatworms (Platyhelminthes: Trematoda). Ph.D. Thesis, Univ. of Leeds.

128. HALTON, D. W. and JENNINGS, J. B. 1964. Demonstration of the nervous system in the monogenetic trematode *Diplozoon paradoxum* Nordmann by the indoxyl acetate method for esterases. *Nature, Lond.*, **202**: 510–511.

128*a*. —1965. Observations on the nutrition of monogenetic trematodes. *Biol. Bull. mar. biol. Lab., Woods Hole*, **129**: 257–272.

129. HENDELBERG, J. 1963. Paired flagella and nucleus migration in the spermiogenesis of *Dicrocoelium* and *Fasciola* (Digenea, Trematoda). *Zool. Bidr. Upps.*, **35**: 569–587.

130. HERBER, E. C. 1950. Studies on the biochemistry of cyst envelopes of the fluke *Notocotylus urbanensis*. *Proc. Pa Acad. Sci.*, **24**: 140–143.

131. HEYNEMAN, D. 1960. On the origin of complex life cycles in the digenetic flukes. *Sobretiro del Libro Homenaje al Dr. Eduardo Caballero y Caballero.* 133–152. (Mexico, 1960.)

132. HOPKINS, C. A. and CALLOW, L. L. 1964. Amino acid exchange between a tapeworm and its environment. *Parasitology*, **54**: 5–6P.

133. HORSTMANN, H. J. 1962. Sauerstoffverbrauch und Glykogengehalt der Eier von *Fasciola hepatica* während der Entwicklung der Miracidien. *Z. ParasitKde.*, **21**: 437–445.

134. HSÜ, H. F., HSÜ, S. Y. LI, and OSBORNE, J. W. 1963. Further studies on Rhesus monkeys immunized against *Schistosoma japonicum* by administration of X-irradiated cercariae. *Z. Tropenmed. Parasit.*, **14**: 402–412.

135. HUANG, T. Y., T'AO, Y. H. and CHU, C. H. 1962. Studies on transaminases of *Schistosoma japonicum*. *Chin. med. J. Peking*, **81**: 79–85.

136. HUFF, C. G. 1940. Immunity in invertebrates. *Physiol. Rev.*, **20**: 68–88.

137. HUGHES, D. L. 1962a. Reduction of the pathogenicity of *Fasciola hepatica* in mice by X-irradiation. *Nature, Lond.*, **193**: 1093–1094.

138. —1962b. Observations on the immunology of *Fasciola hepatica* infections in mice and rabbits. *Parasitology*, **52**: 4P.

139. HUMPHREY, J. H. and WHITE, R. G. 1963. *Immunology for Students of Medicine*. Blackwell, Oxford.

140. HUNTER, III, G. W. and CRANDALL, R. B. 1961. Experiments on some factors affecting resistance to *Schistosoma mansoni* infections in mice. *ASB. Bull.*, **8**: 25.

141. HUNTER, III, G. W., CRANDALL, R. B., ZICKAFOOSE, D. E. and PURVIS, Q. B. 1962. Studies on schistosomiasis. XVIII. Some factors affecting resistance to *Schistosoma mansoni* infections in albino mice. *Am. J. trop. Med. Hyg.*, **11**: 17–24.

142. HUNTER, III, G. W. and HAMILTON, J. M. 1941. Studies on host-parasite reactions to larval parasites. IV. The cyst of *Uvulifer ambloplitis* (Hughes). *Trans. Am. microsc. Soc.*, **60**: 498–507.

143. HUNTER, III, G. W., WEINMANN, C. J. and HOFFMANN, R. G. 1961. Studies on Schistosomiasis. XVII. Non-reciprocal acquired resistance between *Schistosoma mansoni* and *Schistosomatium douthitti* in mice. *Expl Parasit.*, **11**: 133–140.

144. HUNTER, W. S. and VERNBERG, W. B. 1955b. Studies on oxygen consumption in digenetic trematodes. II. Effects of two extremes in oxygen tension. *Expl Parasit.*, **4**; 427–434.

145. HURST, C. T. and WALKER, C. R. 1935. Increased heat production in a poikilothermous animal in parasitism. *Am. Nat.*, **69**: 461–466.

146. HYMAN, L. H. 1951. *The Invertebrates*, Vol. II. McGraw-Hill Book Co., Inc., N.Y.

147. JENNINGS, J. B. 1956. A technique for the detection of *Polystoma integerrimum* in the common frog. *J. Helminth.*, **30**: 119–120.

148. —1959. Studies on digestion in the monogenetic trematode *Polystoma integerrimum*. *J. Helminth.*, **33**: 197–204.

149. —1963. Some aspects of nutrition in the Turbellaria, Trematoda, and Rhynchocoela. In: *The Lower Metazoa*, ed. Dougherty, E. C., Brown, Z. N., Hanson, E. D., and Hartman, W. D. Univ. Calif. Press, Berkeley. Chapter 26.

150. JOHRI, L. N. and SMYTH, J. D. 1956. A histochemical approach to the study of helminth morphology. *Parasitology*, **46**: 107–116.

151. JONES, T. W. T. 1950. Anopheline larvae as intermediate hosts for larval trematodes. *Parasitology*, **40**: 144–148.

152. KABAT, E. A. and MAYER, M. M. 1961. *Experimental Immunochemistry*. 2nd ed. Thomas, Illinois.

153. KAGAN, I. G. and NORMAN, L. 1963. Analysis of helminth antigens. (*Echinococcus granulosus* and *Schistosoma mansoni*) by agar gel methods. *Ann. N.Y. Acad. Sci.*, **113**: 130–153.

154. KAWASHIMA, K., TADA, I. and MIYAZAKI, I. 1961. Host preference of miracidia of *Paragonimus ohirai* Miyazaki, 1939, among three species of snails of the genus *Assiminea*. *Kyushu J. med. Sci.*, **12**: 99–106.

155. KEARN, G. C. 1962. Breathing movements in *Entobdella soleae* (Trematoda, Monogenea) from the skin of the common sole. *J. mar. biol. Ass. U.K.*, **42**: 93–104.

156. —1963a. The life cycle of the monogenean *Entobdella soleae*, a skin parasite of the common sole. *Parasitology*, **53**: 253–263.

157. —1963b. Feeding in some monogenean skin parasites: *Entobdella soleae* on *Solea solea* and *Acanthocotyle* sp. on *Raia clavata*. *J. mar. biol. Ass. U.K.*, **43**: 749–766.

158. —1964. The attachment of the monogenean *Entobdella soleae* to the skin of the common sole. *Parasitology*, **54**: 327–336.

159. KENDALL, S. B. and McCULLOUGH, F. S. 1951. The emergence of the cercariae of *Fasciola hepatica* in *Lymnaea truncatula*. *J. Helminth.*, **25**: 77–92.

160. KENT, J. F. 1963a. Immunodiagnosis of helminthic infections. VI. Current and potential value of immunodiagnostic tests employing soluble antigens. *Am. J. Hyg. Monographic Ser.*, **22**: 68–90.

161. —1963b. Immunodiagnosis of helminthic infections. III. Fractionation, isolation and definition of antigens from parasitic helminths. *Am. J. Hyg. Monographic Ser.*, **22**: 30–45.

162. KIRAKINI, V. V. 1961. [The epizootiology of fascioliasis in sheep in the Turkmen S.S.R.] *Conf. Nat. Focal Occur. of Dis. & Prob. in Parasit. of Kazakhstan & Central Asian Republics (4th) Alma-Ata*, 1959, **3**: 389–393. (In Russian.)

163. KLOETZEL, K. 1958. Observações sôbre o tropismo de miracidio de *Schistosoma mansoni* pelo molusco *Australorbis glabratus*. *Revta bras. Biol.*, **18**: 223–232.

164. KOHLMANN, F. W. 1961. Untersuchungen zur Biologie, Anatomie, und Histologie von *Polystomum integerrimum* Fröhlich. *Z. ParasitKde.*, **20**: 495–524.

165. KOLMOGROVA, E. J. 1959. [Structure of the central parts of the nervous system of *Opisthorchis felineus*.] *Zool. Zh.*, **38**: 1627–1633. (In Russian.)

166. KRUIDENIER, F. J. 1951. The formation and function of mucoids in virgulate cercariae, including a study of the virgula organ. *Am. Midl. Nat.*, **46**: 660–683.

167. —1953. Studies on mucoid secretion and function in the cercaria of *Paragonimus kellicotti* Ward (Trematoda: Troglotrematidae). *J. Morph.*, **92**: 531–538.

168. KRULL, W. H. 1935. Studies on the life history of *Halipegus occidualis* Stafford, 1905. *Am. Midl. Nat.*, **16**: 129–143.

169. KRULL, W. H. and MAPES, C. R. 1951–56. Studies on the biology of *Dicrocoelium dendriticum* (Rudolphi, 1819) Looss, 1899 (Trematoda: Dicrocoeliidae), including its relation to the intermediate host, *Cionella lubrica* (Müller). I–IX. *Cornell Vet.*, Vols. 41–45.

170. KÜMMEL, G. 1958. Das Terminalorgan der Protonephridien, Feinstrucktur und Deutung der Funktion. *Z. Naturf.*, B, **13**: 677–679.

171. KURELEC, B. and EHRLICH, I. 1963. Über die Natur der von *Fasciola hepatica* (L.) *in vitro* ausgeschiedenen Amino- und Ketosäuren. *Expl Parasit.*, **13**: 113–117.

172. LAZARUS, M. 1950. The respiratory metabolism of helminths. *Aust. J. Scient. Res.*, **B3**: 245–250.

173. LEE, C. L. and LEWERT, R. M. 1957. Studies on the presence of mucopolysaccharidase in penetrating helminth larvae. *J. infect. Dis.*, **101**: 287–294.

174. LEE, D. L. 1962. Studies on the function of the pseudosuckers and holdfast organ of *Diplostomum phoxini* Faust (Strigeida, Trematoda). *Parasitology*, **52**: 103–112.

175. LEE, H. F. 1962. Life history of *Heterobilharzia americana* Price 1929, a schistosome of the raccoon and other mammals in southeastern United States. *J. Parasit.*, **48**: 728–739.

176. LEES, E. 1962. The incidence of helminth parasites in a particular frog population. *Parasitology*, **52**: 95–102.

177. LEES, E. and BASS, L. 1960. Sex hormones as a possible factor influencing the level of parasitization in frogs. *Nature, Lond.*, **188**: 1207–1208.

178. LEES, E. and MITCHELL, J. B. 1964. The development of *Gorgoderina vitelliloba* in the definitive host. *Parasitology*, **54**: 14–15P.

179. LEFLORE, W. B. 1961. Serology of marine trematodes. I. The action of antisera against strigeoid and schistosome cercariae. *J. Parasit.*, **47**: 899–904.

180. LENGY, J. 1962. Studies on *Schistosoma bovis* (Sonsino, 1876) in Israel. I. Larval stages from egg to cercaria. *Bull. Res. Coun. Israel*, **10E**: 1–36.

181. LENHOFF, H. M., SCHROEDER, R. and LEIGH, W. H. 1960. The collagen-like nature of metacercarial cysts of a new species of *Ascocotyle*. *J. Parasit.*, **46**: 5, suppl. 36.

182. LEVINE, D. M. and KAGAN, I. G. 1960. Studies on the immunology of schistosomiasis by vaccination and passive transfer. *J. Parasit.*, **46**: 787–792.

183. LEWERT, R. M. and LEE, C. L. 1954. Studies on the passage of helminth larvae through host tissues. I. Histochemical studies on the extracellular changes caused by penetrating larvae. II. Enzymatic activity of larvae *in vitro* and *in vivo*. *J. infect. Dis.*, **95**: 13–51.

184. —1956. Quantitative studies of the collagenase-like enzymes of cercariae of *Schistosoma mansoni* and the larvae of *Strongyloides ratti*. *J. infect. Dis.*, **99**: 1–14.

185. LIN, S. S. and SADUN, E. H. 1959. Studies on the host-parasite relationships to *Schistosoma japonicum*. V. Reactions in the skin, lungs, and liver of normal and immune animals following infection with *Schistosoma japonicum*. *J. Parasit.*, **45**: 549–559.

186. LINCICOME, D. R. ed. 1962. Frontiers in research in parasitism. I. Cellular and humoral reactions in experimental schistosomiasis. *Expl Parasit.*, **12**: 211–240.

187. LLEWELLYN, J. 1954. Observations on the food and the gut pigment of the *Polyopisthocotylea* (Trematoda: Monogenea). *Parasitology*, **44**: 428–437.

188. —1956. The host specificity, microecology, adhesive attitudes and comparative morphology of some trematode gill parasites. *J. mar. biol. Ass. U.K.*, **35**: 113–127.

189. —1957*a*. The mechanism of the attachment of *Kuhnia scombri* (Kuhn, 1829) (Trematoda: Monogenea) to the gills of its host *Scomber scombrus* L. including a note on the taxonomy of the parasite. *Parasitology*, **47**: 30–39.

190. —1957*b*. The larvae of some monogenetic trematode parasites of Plymouth fishes. *J. mar. biol. Ass. U.K.*, **36**; 243–259.

191. —1957*c*. Host-specificity in monogenetic trematodes. In: *First symposium on host specificity among parasites of vertebrates, Neuchâtel.* Univ. of Neuchâtel, 199–211.

192. —1958. The adhesive mechanisms of monogenetic trematodes; the attachment of species of the Diclidophoridae to the gills of gadoid fishes. *J. mar. biol. Ass. U.K.*, **37**: 67–79.

193. —1962*a*. The life histories and population dynamics of monogenean gill parasites of *Trachurus trachurus* (L.). *J. mar. biol. Ass. U.K.*, **42**: 587–600.

194. —1962*b*. The effects of the host and its habits on the morphology and life-cycle of a monogenean parasite. In: *Symposium on Helminths Bound to Aquatic Conditions* (1962), Prague: Parasitologicky Ustav Ceskoslovenske Akademie Ved.

195. —1963. Larvae and larval development of Monogeneans. In: *Advances in Parasitology* (ed. B. Dawes), **1**: 287–326.

196. LLEWELLYN, J. and EUZET, L. 1964. Spermatophores in the monogenean *Entobdella diadema* Monticelli from the skin of sting-rays, with a note on the taxonomy of the parasite. *Parasitology*, **54**: 337–344.

197. LLEWELLYN, J. and OWEN, I. L. 1960. The attachment of the monogenean *Discocotyle sagittata* Leuckart to the gills of *Salmo trutta* L. *Parasitology*, **50**: 51–60.

198. LYNCH, D. L. and BOGITSH, B. J. 1962. The chemical nature of metacercarial cysts. II. Biochemical investigations on the cyst of *Posthodiplostomum minimum*. *J. Parasit.*, **48**: 241–243.

199. LYONS, K. M. 1964. The chemical nature and evolutionary significance of monogenean attachment sclerites. *Parasitology*, **54**: 12P.

200. MA, L. 1963. Trace elements and polyphenol oxidase in *Clonorchis sinensis*. *J. Parasit.*, **49**: 197–203.

201. —1964. Acid phosphatase in *Clonorchis sinensis*. *J. Parasit.*, **50**: 235–240.

202. MACFARLANE, W. V. 1949*a*. Schistosome dermatitis in New Zealand. Pt. I. The parasite. *Am. J. Hyg.*, **50**: 143–151.

203. —1949*b*. Schistosome dermatitis in New Zealand. Pt. II. Pathology and immunology of cercarial lesions. *Am. J. Hyg.*, **50**: 152–167.

204. MACINNIS, A. J. 1963*a*. Responses of *Schistosoma mansoni* miracidia to chemical attractants. *J. Parasit.*, **49** (suppl.), 19.

205.—1963*b*. Evidence for chemical attraction of *Schistosoma mansoni* miracidia. *J. Parasit.*, **49**: (suppl.), 19.

206. MANDLOWITZ, S., DUSANIC, D. and LEWERT, R. M. 1960. Peptidase and lipase activity of extracts of *Schistosoma mansoni* cercariae. *J. Parasit.*, **46**: 89–90.

207. MANSOUR, T. E. 1957*a*. The effect of lysergic acid diethylamide, 5-hydroxytryptamine, and related compounds on the liver fluke, *Fasciola hepatica*. *Br. J. Pharmac. Chemother.*, **12**: 406–409.

208. —1957*b*. The effect of serotonin and related amines on the activity of phenol oxidases. *J. Pharmac. exp. Ther.*, **119**: 164.

209. —1958. Effect of serotonin on phenol oxidase from the liver fluke *Fasciola hepatica* and from other sources. *Biochim. biophys. Acta*, **30**: 492–500.

210. —1959a. The effect of serotonin and related compounds on the carbohydrate metabolism of the liver fluke, *Fasciola hepatica*. *J. Pharmac. exp. Ther.*, **126**: 212–216.

211. —1959b. Studies on the carbohydrate metabolism of the liver fluke *Fasciola hepatica*. *Biochim. biophys. Acta*, **34**: 456–464.

212. —1962. Effect of serotonin on glycolysis in homogenates from the liver fluke *Fasciola hepatica*. *J. Pharmac. exp. Ther.*, **135**: 94–101.

213. MANSOUR, T. E. and BUEDING, E. 1954. The actions of antimonials on glycolytic enzymes of *Schistosoma mansoni*. *Br. J. Pharmac. Chemother.*, **9**: 459–462.

214. MANSOUR, T. E., BUEDING, E. and STAVITSKY, A. B. 1954. The effect of a specific antiserum on the activities of lactic dehydrogenase of mammalian muscle and of *Schistosoma mansoni*. *Br. J. Pharmac. Chemother.*, **9**: 182–186.

215. MANSOUR, T. E. and MANSOUR, J. M. 1962. Effects of serotonin (5-hydroxytryptamine) and adenosine 3′, 5′-phosphate on phosphofructokinase from the liver fluke *Fasciola hepatica*. *J. biol. Chem.*, **237**: 629–634.

216. MANSOUR, T. E., SUTHERLAND, E. W., RALL, T. W. and BUEDING, E. 1960. The effect of serotonin (5-hydroxytryptamine) on the formation of adenosine 3′, 5′-phosphate by tissue particles from the liver fluke *Fasciola hepatica*. *J. biol. Chem.*, **235**: 466–470.

217. MATTES, O. 1949. Wirtsfindung und wirtsspezificität bei parasiten. *Verh. dt. zool. Ges.*, 1948, 165–173.

218. MEISENHELDER, J. E., OLSZEWSKI, B. and THOMPSON, P. E. 1960. Observations on therapeutic and prophylactic effects by homologous immune blood against *Schistosoma mansoni* in rhesus monkeys. *J. Parasit.*, **46**: 645–647.

219. MELEKH, D. A. 1963. [Comparative study of glycogen reserves in some ecto- and endoparasites.] *Vest. leningr. gos. Univ., Ser. Biol.*, **18**: 5–13. (In Russian.)

220. MILLEMANN, R. E. and THONARD, J. C. 1959. Protease activity in schistosome cercariae. *Expl Parasit.*, **8**: 129–136.

221. MIRETSKI, O. Y. 1951. Experiment on controlling the process of vital activity of the helminth by influencing the condition of the host. *Dokl. obshchem. Sobran Akad. Nauk USSR*, **78**: 613–615.

222. MIURA, M. 1955. [Histochemical studies on *Schistosoma japonicum*. 3. Histochemistry of the cercaria of *Schistosoma japonicum*.] *Kumamoto Igakkwai Zasshi*, **29**: 99–109. (In Japanese.)

223. MOFTY, M. M. EL and SMYTH, J. D. 1964. Endocrine control of encystation in *Opalina ranarum* parasitic in *Rana temporaria*. *Expl Parasit.*, **25**: 185–199.

224. MOORE, D. V., YOLLES, T. K. and MELENEY, H. E. 1954. The relationship of male worms to the sexual development of female *Schistosoma mansoni*. *J. Parasit.*, **40**: 166–185.

225. MOROZOV, F. N. 1958. [The question of the presence of an anus in digenetic trematodes.] (Papers on Helminthology presented to Academician K. I. Skryabin on his 80th birthday). *Izdatelstvo Akad. Nauk SSSR*, 239–242. (In Russian.)

226. MUELLER, J. F. 1963. Parasite-induced weight gain in mice. *Ann. N. Y. Acad. Sci.*, **113**: 217–233.

227. NADAKAL, A. M. 1960. Chemical nature of cercarial eye-spot and other tissue pigments. *J. Parasit.*, **46**: 475–483.

228. NEWSOME, J. 1962. Maturation of schistosome eggs *in vitro*. *Nature, Lond.*, **195**: 722–723.

229. NEWSOME, J. and ROBINSON, D. L. H. 1954. Investigation of methods of maintaining *Schistosoma mansoni in vitro*. *Ann. trop. Med. Parasit.*, **48**: 194–200.

230. NEWTON, W. L. 1952. The comparative tissue reaction of two strains of *Australorbis glabratus* to infection with *Schistosoma mansoni*. *J. Parasit.*, **38**: 362–366.

231. —1953. The inheritance of susceptibility to infection with *Schistosoma mansoni* in *Australorbis glabratus*. *Expl Parasit.*, **2**: 242–257.

232. —1954. Tissue response to *Schistosoma mansoni* in second generation snails from a cross between two strains of *Australorbis glabratus*. *J. Parasit.*, **40**: 352–355.

233. NIMMO-SMITH, R. H. and STANDEN, O. D. 1963. Phosphomonoesterases of *Schistosoma mansoni*. *Expl Parasit.*, **13**: 305–322.

234. OGLESBY, L. C. 1961. Ovoviviparity in the monogenetic trematode *Polystomoidella oblonga*. *J. Parasit.*, **47**: 237–243.

235. OLIVIER, L., VON BRAND, T. and MEHLMAN, B. 1953. The influence of lack of oxygen on *Schistosoma mansoni* cercariae and on infected *Australorbis glabratus*. *Expl Parasit.*, **2**: 258–270.

236. ORTIGOZA, R. O. and HALL, J. E. 1963a. Studies on the glandular apparatus and secretions of virgulate xiphidiocercariae. I. Intravital and histochemical data. *Expl Parasit.*, **14**: 160–177.

237. —1963b. Studies on the glandular apparatus and secretions of virgulate xiphidiocercariae. II. Qualitative chemical analysis of secretions. *Expl Parasit.*, **14**: 178–185.

238. OSHIMA, T., YOSHIDA, Y. and KIHATA, M. 1958. [Studies on the excystation of the metacercariae of *Paragonimus westermani*. 1. Especially on the effect of bile salts.] *Bull. Inst. publ. Hlth., Tokyo*, **7**: 256–269. (In Japanese.)

239. PANTELOURIS, E. M. and HALE, P. A. 1962. Iron and vitamin C in *Fasciola hepatica* L. *Res. vet. Sci.*, **3**: 300–303.

240. PEARSON, I. G. 1963. Use of chromium radioisotope ^{51}Cr to estimate blood loss through ingestion by *Fasciola hepatica*. *Expl Parasit.*, **13**: 186–193.

241. PEARSON, J. C. 1956. Studies on the life cycles and morphology of the larval stages of *Alaria arisaemoides* Augustine and Uribe, 1927, and *Alaria canis* LaRue and Fallis, 1936 (Trematoda: Diplostomidae). *Can. J. Zool.*, **34**: 295–387.

242. —1959. Observations on the morphology and life cycle of *Strigea elegans* Chandler and Rausch, 1947 (Trematoda; Strigeidae). *J. Parasit.*, **45**: 155–174.

243. —1961. Observations on the morphology and life cycle of *Neodiplostomum intermedium* (Trematoda: Diplostomatidae). *Parasitology*, **51**: 133–172.

244. —1965. Unpublished work.

245. PENNOIT-DE-COOMAN, E. and VAN GREMBERGEN, G. 1942. Vergelijkend Onderzoek van het fermentensysteem bij vrijlevende en parasitaire Plathelminthen. *Verh. K. vlaam. Acad. Wet.*, **4**: 7–77.

246. PEPLER, W. J. 1958. Histochemical demonstration of an acetylcholin-esterase in the ova of *Schistosoma mansoni*. *J. Histochem. Cytochem.*, **6**: 139–141.

247. PROST, M. 1959*a*. Wpływ czynników ekologicznych na faune Mono-genoidea u ryb. *Wiad. parazyt.*, **5**: 453–458.

248. —1959*b*. Badania nad wpływem zasolenia wody na faune Monogenoidea ryb. *Acta parasit. pol.*, **7**: 615–630.

249. RADKE, M. G., RITCHIE, L. S. and ROWAN, W. B. 1961. Effects of water velocities on worm burdens of animals exposed to *Schistosoma mansoni* cercariae released under laboratory and field conditions. *Expl Parasit.*, 11: 323–331.

250. RAI, S. L. 1965. Unpublished work.

251. READ, C. P. 1961. The carbohydrate metabolism of worms. In: *Comparative Physiology of Carbohydrate Metabolism in Heterothermic Animals.* Univ. Wash. Press.

252. READ, C. P., ROTHMAN, A. H. and SIMMONS, J. E., JR. 1963. Studies on membrane transport, with special reference to parasite-host integration. *Ann. N.Y. Acad. Sci.*, **113**: 154–205.

253. READ, C. P. and YOGORE, M. 1955. Respiration of *Paragonimus wester-mani*. *J. Parasit.*, **41**, 6 Sect. 2, 28.

254. REES, F. G. 1934. *Cercaria patellae* Lebour, 1911 and its effects on the digestive gland and gonads of *Patella vulgata*. *Proc. zool. Soc. Lond.*, 45–53.

255. —1939. Studies on the germ-cycle of the digenetic trematode, *Parorchis acanthus* Nicoll. Part I. Anatomy of the genitalia and gametogenesis in the adult. *Parasitology*, **31**: 417–433.

256. —1955. The adult and diplostomulum stage (*Diplostomulum phoxini* (Faust)) of *Diplostomum pelmatoides* Dubois, and an experimental demonstration of part of the life cycle. *Parasitology*, **45**: 295–312.

257. REES, W. J. 1936. The effect of parasitism by larval trematodes on the tissues of *Littorina littorea* (Linné). *Proc. zool. Soc. Lond.*, 357–368.

258. REISINGER, E. and GRAAK, B. 1962. Untersuchungen an *Codonocephalus* (Trematoda Digenea: Strigeidae), Nervensystem und Paranephridi-aler plexus. *Z. ParasitKde.*, **22**: 1–42.

259. ROBERTS, E. W. 1950. Studies on the life-cycle of *Fasciola hepatica* (Linnaeus) and of its snail host, *Lymnaea (Galba) truncatula* (Müller), in the field and under controlled conditions in the laboratory. *Ann. trop. Med. Parasit.*, **44**: 187–206.

260. ROBINSON, D. L. H. 1956. A routine method for the maintenance of *Schistosoma mansoni in vitro*. *J. Helminth.*, **29**: 193–202.

261. —1960. Egg-laying by *Schistosoma mansoni in vitro*. *Ann. trop. Med. Parasit.*, **54**: 112–117.

262. —1961. Phosphatases in *Schistosoma mansoni*. *Nature, Lond.*, **191**: 473–474.

263. ROGERS, W. P. 1940. Haematological studies on the gut contents of certain nematode and trematode parasites. *J. Helminth.*, **18**: 53–62.

264. —1949. On the relative importance of aerobic metabolism in small nematode parasites of the alimentary tract. I. Oxygen tensions in the normal environment of the parasites. *Aust. J. scient. Res.*, **2B**. 157–165.

265. —1962. *The Nature of Parasitism.* Academic Press, London and New York.

266. ROWAN, W. B. 1956. The mode of hatching of the egg of *Fasciola hepatica. Expl Parasit.,* **5**: 118–137.

267. 1957. The mode of hatching of the egg of *Fasciola hepatica.* II. Colloidal nature of the viscous cushion. *Expl Parasit.,* **6**: 131–142.

268. ROWCLIFFE, S. A. and OLLERENSHAW, C. B. 1960. Observations on the bionomics of the egg of *Fasciola hepatica. Ann. trop. Med. Parasit.,* **54**: 172–181.

269. RUSZOWSKI, J. S. 1931. Sur la découverte d'un ectoparasite *Amphibdella torpedinis* dans la cœur des torpilles. *Pubbl. Staz. zool. Napoli,* **11**: 161–167.

270. SADUN, E. H. 1963. Immunization in schistosomiasis by previous exposure to homologous and heterologous cercariae by inoculation of preparations from schistosomes and by exposure to irradiated cercariae. *Ann. N.Y. Acad. Sci.,* **113**: 418–439.

271. SADUN, E. H. and LIN, S. S. 1959. Studies on the host-parasite relationships to *Schistosoma japonicum.* IV. Resistance acquired by infection, by vaccination and by the injection of immune serum, in monkeys, rabbits and mice. *J. Parasit.,* **45**: 543–548.

272. SADUN, E. H., WALTON, B. C., BUCK, A. A. and LEE, B. K. 1959. The use of purified antigens in the diagnosis of clonorchiasis *sinensis* by means of intradermal and complement fixation tests. *J. Parasit.,* **45**: 129–134.

273. SAITO, A. 1961. [Histochemical study on the digestive tract of *Fasciola hepatica.*] *J. Tokyo med. Coll.,* **19**: 1487–1497. (In Japanese.)

274. SALT, G. 1963. The defence reactions of insects to metazoan parasites. *Parasitology,* **53**: 527–642.

275. SCHELL, S. C. 1962. Development of the sporocyst generations of *Glypthelmins quieta* (Stafford, 1900) (Trematoda: Plagiorchioidea), a parasite of frogs. *J. Parasit.,* **48**: 387–394.

276. SEITNER, P. G. 1945. Studies on five new species of xiphidiocercariae of the virgula type. *J. Parasit.,* **31**: 272–281.

277. SENFT, A. W. 1963. Observations on amino acid metabolism of *Schistosoma mansoni* in a chemically defined medium. *Ann. N.Y. Acad. Sci.,* **113**: 272–288.

278. —1964. Recent developments in understanding of amino acid and protein metabolism *in vitro* by *Schistosoma mansoni. Proc. 1st Int. Congr. Parasitology,* Rome, 1964.

279. SENFT, A. W., PHILPOTT, D. E. and PELOFSKY, A. H. 1961. Electron microscope observations of the integument, flame cells, and gut of *Schistosoma mansoni. J. Parasit.,* **47**: 217–229.

280. SENFT, A. W. and SENFT, D. G. 1962. A chemically defined medium for maintenance of *Schistosoma mansoni. J. Parasit.,* **48**: 551–554.

281. SENTERFIT, L. B. 1957. The development of antibodies in schistosome infections as measured by the miracidial immobilization reaction. *Am. J. trop. Med. Hyg.,* **6**: 390–391.

282. SHAPIRO, J. E. 1961. The structure and development of the cuticulum of digenetic trematodes. *Diss. Abstr.,* **21**: 3239.

283. SHAPIRO, J. E., HERSHENOV, B. R. and TULLOCH, G. S. 1961. The fine structure of *Haematoloechus* spermatozoan tail. *J. biophys. biochem. Cytol.,* **9**: 211–217.

284. SHERIF, A. F. 1962. Schistosomal metabolic products in the diagnosis of bilharziasis. In: *Bilharziasis*. *Ciba Foundation Symposium*, 226–238. Churchill, London.

285. SINGH, K. S. and LEWERT, R. M. 1959. Observations on the formation and chemical nature of metacercarial cysts of *Notocotylus urbanensis*. *J. infect. Dis.*, **104**: 138–141.

286. SMIT, G. J. 1961. The ' Cercarien Hüllen Reaktion ' and the cercaricidal reaction. *Trop. geogr. Med.*, **13**: 374–377.

287. SMITHERS, S. R. 1962a. Acquired resistance to bilharziasis. In: *Bilharziasis*. *Ciba Foundation Symposium*. 239–258. Churchill, London.

288. ——1962b. Stimulation of acquired resistance to *Schistosoma mansoni* in monkeys: role of eggs and worms. *Expl Parasit.*, **12**: 263–273.

289. SMYTH, D. H. 1963. Intestinal absorption. In : *Recent Advances in Physiology* (8th edition). Churchill, London.

290. SMYTH, J. D. 1962a. *Introduction to Animal Parasitology*. English University Press, London.

291. ——1962b. Lysis of *Echinococcus granulosus* by surface-active agents in bile and the role of this phenomenon in determining host specificity in helminths. *Proc. R. Soc.*, B, **156**: 553–572.

292. SMYTH, J. D. and CLEGG, J. A. 1959. Egg-shell formation in trematodes and cestodes. *Expl Parasit.*, **8**: 286–323.

293. SMYTH, J. D. and HASLEWOOD, G. A. D. 1963. The biochemistry of bile as a factor in determining host specificity in intestinal parasites, with particular reference to *Echinococcus granulosus*. *Ann. N.Y. Acad. Sci.*, **113**: 234–260.

294. SOULSBY, E. J. 1962. Antigen-antibody reactions in helminth infections. *Adv. Immun.*, **2**: 265–308.

295. ——1962. Immunity to helminths and its effect on helminth infections. In: *Animal Health and Production*. (Eds. C. S. Grunsell, and A. I. Wright) Butterworths, London.

296. SPRENT, J. F. A. 1963. *Parasitism*. University of Queensland Press, Brisbane.

297. STANDEN, O. D. 1953a. The penetration of the cercariae of *Schistosoma mansoni* into the skin and lymphatics of the mouse. *Trans. R. Soc. trop. Med. Hyg.*, **47**: 292–298.

298. ——1953b. Experimental schistosomiasis. III. Chemotherapy and mode of drug action. *Ann. trop. Med. Parasit.*, **47**: 26–43.

299. STANDEN, O. D. and FULLER, K. A. 1959. Ultra-violet irradiation of the cercariae of *Schistosoma mansoni*. Inhibition of development to the adult stage. *Trans. R. Soc. trop. Med. Hyg.*, **53**: 372–379.

300. STAUBER, L. A. 1961. Immunity in invertebrates with special reference to the oyster. *Proc. natn. Shellfish. Ass.*, **50**: 7–20.

301. STEPHENSON, W. 1947. Physiological and histochemical observations on the adult liver fluke, *Fasciola hepatica* L. I–IV. *Parasitology*, **38**: 116–144.

302. STIREWALT, M. A. 1958. Relation of skin reaction to penetration and to the development of local resistance to entry by challenging cercariae of *Schistosoma mansoni*. *Proc. 6th Int. Congr. trop. Med. Malar.*, II, 67–76.

303. ——1959. Chronological analysis, pattern and rate of migration of cercariae of *Schistosoma mansoni* in body, ear and tail skin of mice. *Ann. trop. Med. Parasit.*, **53**: 400–413.

304. —1963a. Cercaria vs. Schistosomule (*Schistosoma mansoni*): Absence of the pericercarial envelope *in vivo* and the early physiological and histological metamorphosis of the parasite. *Expl Parasit.*, **13**: 395–406.

305. —1963b. Seminar on immunity to parasitic helminths. IV. Schistosome infections. *Expl Parasit.*, **13**: 18–44.

306. —1963c. Chemical biology of secretions of larval helminths. *Ann. N.Y. Acad. Sci.*, **113**: 36–53.

307. STIREWALT, M. A. and EVANS, A. S. 1952. Demonstration of an enzymatic factor in cercariae of *Schistosoma mansoni* by the streptococcal decapsulation test. *J. infect. Dis.*, **91**: 191–197.

308. —1955. Serologic reactions in *Schistosoma mansoni* infections. I. Cercaricidal, precipitation, agglutination, and CHR phenomena. *Expl Parasit.*, **4**: 123–142.

309. —1960. Chromatographic analysis of secretions from the acetabular glands of cercariae of *Schistosoma mansoni*. *Expl Parasit.*, **10**: 75–80.

310. STIREWALT, M. A. and KRUIDENIER, F. J. 1961. Activity of the acetabular secretory apparatus of cercariae of *Schistosoma mansoni* under experimental conditions. *Expl Parasit.*, **11**: 191–211.

311. STUNKARD, H. W. 1931. Further observations on the occurrence of anal openings in digenetic trematodes. *Z. ParasitKde.*, **3**: 713–725.

312. —1955. Induced gametogenesis in a monogenetic trematode, *Polystoma stellai* Vigueras, 1955. *J. Parasit.*, **45**: 389–394.

313. STYCZYŃSKA-JUREWICZ, E. 1961. On the geotaxis, invasivity and span of life of *Opisthioglyphe ranae* Duj. cercariae. *Bull. Acad. pol. Sci. Cl. II. Sér. Sci. biol.*, **9**: 31–35.

314. SUDDS, JR. R. H. 1960. Observations of schistosome miracidial behavior in the presence of normal and abnormal snail hosts and subsequent tissue studies of these hosts. *J. Elisha Mitchell scient. Soc.*, **76**: 121–133.

315. TAKAHASHI, T., MORI, K. and SHIGETA, Y. 1961. Phototactic, thermotactic and geotactic responses of miracidia of *Schistosoma japonicum*. *Jap. J. Parasit.*, **10**: 686–691.

316. TANDON, R. S. 1960. Studies on the lymphatic system of amphistomes of ruminants: I. *Carmyerius spatiosus* (Stiles and Goldberger, 1910). *Zool. Anz.*, **164**: 213–217.

317. THORPE, E. and BROOME, A. W. J. 1962. Immunity to *Fasciola hepatica* infection in albino rats vaccinated with irradiated metacercariae. *Vet. Rec.*, **74**: 755–756.

318. THREADGOLD, L. T. 1963. The tegument and associated structures of *Fasciola hepatica*. *Q. Jl. microsc. Sci.*, **104**: 505–512.

319. 1965. Personal communication.

320. THURSTON, J. P. 1964. The morphology and life cycle of *Protopolystoma xenopi* (Price) Bychowsky in Uganda. *Parasitology*, **54**: 441–450.

321. TIMMS, A. R. 1960. Schistosome enzymes. In: *Host Influence on Parasite Physiology*. Rutgers University Press, N.J.

322. TIMMS, A. R. and BUEDING, E. 1959. Studies of a proteolytic enzyme from *Schistosoma mansoni*. *Br. J. pharmac. Chemother.*, **14**: 68–73.

323. TIMON-DAVID, J. 1958. Rôle des insectes comme hôtes intermédiaires dans les cycles des trématodes digénétiques. *Proc. 10th Int. Congr. Ent.* (1956), **3**: 657–662.

324. —1960. Recherches expérimentales sur le cycle de *Dicrocoelioides petiolatum* (A. Raillet, 1900) (Trematoda, Dicrocoeliidae). *Annls Parasit. hum. comp.*, **35**: 251–267.

325. —1961. Rôle du têtard de *Rana esculenta* L. comme second hôte intermédiaire dans le cycle d'*Ascocotyle branchialis* nov. sp. (Trematoda, Digenea, Heterophyidae). *C.r. hebd. Séanc. Acad. Sci., Paris*, **252**: 3122–3124.

326. TRIPP, M. R. 1963. Cellular responses to mollusks. *Ann. N.Y. Acad. Sci.*, **113**: 467–474.

327. TULLOCH, G. S., SHAPIRO, J. E. and HERSHENOV, B. 1961. Protoplasmic processes of epithelial cells of *Gorgodera*. *J. Parasit.*, **47**: 509–510.

328. UDE, J. 1962. Neurosekretorische Zellen im Cerebralganglion von *Dicrocoelium lanceatum* St. u. H. (Trematoda-Digenea). *Zool. Anz.*, **169**: 455–457.

329. USPENSKAYA, A. V. 1962. [Nutrition of monogenetic trematodes.] *Dokl. Akad. Nauk SSSR*, **142**: 1212–1215. (In Russian.)

330. VAN CLEAVE, H. J. and WILLIAMS, C. O. 1943. Maintenance of a trematode, *Aspidogaster conchicola*, outside of the body of its natural host. *J. Parasit.*, **29**: 127–130.

331. VAN GREMBERGEN, G. 1949. Le métabolisme respiratoire du trématode *Fasciola hepatica* (Linn., 1758). *Enzymologia*, **13**: 241–257.

332. VARMA, A. K. 1961. Observations on the biology and pathogenicity of *Cotylophoron cotylophorum* (Fischoeder, 1901). *J. Helminth.*, **35**: 161–168.

333. VASILEVA, I. N. 1960. [A study of ontogenesis in the trematode *Fasciola hepatica* in the Moscow region.] *Zool. Zh.*, **39**: 1478–1484. (In Russian.)

334. VERNBERG, W. B. 1961a. The influence of temperature on metabolic response in larval trematodes. *J. Parasit.*, **47**: 4, Sect. 2, 43.

335. —1961b. Studies on oxygen consumption in digenetic trematodes. VI. The influence of temperature on larval trematodes. *Expl Parasit.*, **11**: 270–275.

336. —1963. Respiration of digenetic trematodes. *Ann. N.Y. Acad. Sci.*, **113**: 261–271.

337. VERNBERG, W. B. and HUNTER, W. S. 1956. Quantitative determinations of the glycogen content of *Gynaecotyla adunca* (Linton, 1905). *Expl Parasit.*, **5**: 441–448.

338. —1959. Studies on oxygen consumption in digenetic trematodes. III. The relationship of body nitrogen to oxygen uptake. *Expl Parasit.*, **8**: 76–82.

339. —1960. Studies on oxygen consumption in digenetic trematodes. IV. Oxidative pathways in the trematode *Gynaecotyla adunca* (Linton, 1905). *Expl Parasit.*, **9**: 42–46.

340. —1961. Studies on oxygen consumption in digenetic trematodes. V. The influence of temperature on three species of adult trematodes. *Expl Parasit.*, **11**: 34–38.

341. —1963. Utilization of certain substrates by larval and adult stages of *Himasthla quissetensis*. *Expl Parasit.*, **14**: 311–315.

342. VOGEL, H. 1941. Über den Einfluss des Geschlechtspartners auf Wachstum und Entwicklung bei *Bilharzia mansoni* und bei Kreuzpaarungen zwischen verschiedenen Bilharzia-Arten. *Zentbl. Bakt. ParasitKde Abt. 1, originale*, **148**: 78–96.

343. VOGEL, H. and MINNING, W. 1949. Hüllenbildung bei Bilharzia-Cercarien im Serum bilharzia-infizierter Tiere und Menschen. *Zentbl. Bakt. ParasitKde Abt.* i, originale, **153**: 91–105.

344. VON BRAND, T. 1952. *Chemical Physiology of Endoparasitic Animals.* Academic Press, N.Y. and London.

345. VON BRAND, T. and FILES, V. S. 1947. Chemical and histological observations on the influence of *Schistosoma mansoni* infection on *Australorbis glabratus. J. Parasit.*, **33**: 476–482.

346. VON BRAND, T. and MERCADO, T. I. 1961. Histochemical glycogen studies on *Fasciola hepatica. J. Parasit.*, **47**: 459–463.

347. VON LICHTENBERG, F., SADUN, E. H. and BRUCE, J. I. 1962. Tissue responses and mechanisms of resistance in schistosomiasis *mansoni* in abnormal hosts. *Am. J. trop. Med. Hyg.*, **11**: 347–356.

348. WAGNER, A. 1959. Stimulation of *Schistosomatium douthitti* cercariae to penetrate their host. *J. Parasit.*, **45**: 4 Sec. 2, 16.

349. WEINLAND, E. and VON BRAND, T. 1926. Beobachtungen an *Fasciola hepatica. Z. vergl. Physiol.*, **4**: 212–285.

350. WHARTON, G. W. 1941. The function of respiratory pigments of turtle parasites. *J. Parasit.*, **27**: 81–87.

351. W.H.O. & U.N.E.S.C.O. 1964. *Immunological Methods.* (Ed. J. F. Ackroyd.) Blackwell, Oxford.

352. WIKERHAUSER, T. 1960. A rapid method for determining the viability of *Fasciola hepatica* metacercariae. *Am. J. vet. Res.*, **21**: 895–897.

353. —1961a. Immunobiologic diagnosis of fascioliasis. II. The *in vitro* action of immune serum on the young parasitic stage of *F. hepatica*—a new precipitin test for fascioliasis. *Vet. Arh.*, **31**: 71–80.

354. —1961b. O utjecaju rendgenskog zračenja na metacerkarije metilja *F. hepatica. Vet. Arh.*, **31**: 229–236.

355. WILBRANDT, W. and ROSENBERG, T. 1961. The concept of carrier transport and its corollaries in pharmacology. *Pharm. Rev., Lond.*, **13**; 109–183.

356. WILLEY, C. H. and GROSS, P. R. 1957. Pigmentation in the foot of *Littorina littorea* as a means of recognition of infection with trematode larvae. *J. Parasit.*, **43**: 324–327.

357. WILLIAMS, H. H. 1959. The anatomy of *Köllikeria filicollis* (Rudolphi, 1819), Cobbold, 1860 (Trematoda: Digenea) showing that the sexes are not entirely separate as hitherto believed. *Parasitology*, **49**: 39–53.

358. WILLIAMS, M. O., HOPKINS, C. A. and WYLLIE, M. R. 1961. The *in vitro* cultivation of strigeid trematodes. III. Yeast as a medium constituent. *Expl Parasit.*, **11**: 121–127.

359. WILSON, T. 1962. *Intestinal Absorption.* Saunders, London.

360. WOOTTON, D. M. 1957. Studies on the life-history of *Allocreadium alloneotenicum* sp. nov. (Allocreadiidae: Trematoda). *Biol. Bull. mar. biol. Lab., Woods Hole*, **113**: 302–315.

361. WRIGHT, C. A. 1959. Host location by trematode miracidia. *Ann. trop. Med. Parasit.*, **53**: 288–292.

362. WYKOFF, D. E. and LEPEŠ, T. J. 1957. Studies on *Clonorchis sinensis.* I. Observation on the route of migration in the definitive host. *Am. J. trop. Med. Hyg.*, **6**: 1061–1065.

363. WYLLIE, M. R., WILLIAMS, M. O. and HOPKINS, C. A. 1960. The *in vitro* cultivation of strigeid trematodes. II. Replacement of a yolk medium. *Expl Parasit.*, **10**: 51–57.

363*a*. YOKOGAWA, M., OSHIMA, T., and KIHATA, M. 1965.   Studies to maintain excysted metacercariae of *Paragonimus westermani in vitro*. *J. Parasit.*, **41** (Sect. 2): 28.

364. YOKOGAWA, M., YOSHIMURA, H., SANO, M., OKURA, T. and TSUJI, M. 1962.   The route of migration of the larva of *Paragonimus westermani* in the final hosts. *J. Parasit.*, **48**: 525–531.

365. ZEUTHEN, E. 1953.   Oxygen uptake as related to body size in organisms. *Q. Rev. Biol.*, **28**: 1–12.

Index

A number in italics refers to an illustration on that page.